PARTICIPANT COMMENTS

This study is exceptional. It provides sound scientific understanding and theological insight into how we respond to God's creation to please and worship God. This is the best treatment of global climate change I have ever read and the best study group on creation care I have participated in.

Steve's teaching is very insightful, and his book is informative. He has backed up his insights with considerable research not often shared with such understandable and clear explanations.

This book makes the science of climate change easier to understand and challenges us to find ways we can make a difference to help reverse climate change.

This study was extremely informative. It made us aware of ways that each of us can contribute to reducing our carbon footprint in the future. Steve's book and discussion sessions opened our eyes and made us realize the critical need to change how we live now, not some future time.

This creation care study made us more aware and conscious of the steps we can take to make a difference (at home and work) and triggered motivation to take more than a few small steps.

Change is difficult. We can reduce, reuse, and recycle. We can start small and do a few things at a time to change what we do and how we live. Steve's book and teaching helped create a space where we could have constructive conversations and respectfully listen to each other's points of view, even if we did not fully agree. Steve's class helped us build connections and learn together how to help take care of the earth.

This book provides valuable statistics, charts, graphs, and other relevant information supported by extensive scientific research. Steve supports his call to action with moral arguments and Biblical teachings. From the time spent preparing for and teaching this class and the research into this book, it is obvious that Steve cares deeply about climate change.

Steve's teaching and class presentation were inspiring and challenging, and they motivated me to continue to find ways to protect and care for our environment and world. The class was very informative. Steve's knowledge of global warming was all-encompassing and provided an excellent foundation for pursuing action to decrease climate change's effects daily.

CLIMATE CHANGE AND THE HEALING OF CREATION

If there can ever be one bowl to hold the nuances and complexities of this super-wicked global issue, Climate Change and the Healing of Creation is such a container. Moving from hard science to fresh theological framing and tender-hearted lament, readers will complete the 13-week study guide with a clear-sighted and holy call to action. This extensively researched and comprehensive guide is what we need to face the climate crisis bravely, together.

Kirstin De Mello
Mennonite Central Committee U.S. climate advocacy and education coordinator

Dr. Pardini blends careful science and faithful discipleship in this engaging, honest, and passionate manual on how Christian communities can think and act to protect the environment and care for God's good creation. I'd recommend it for every church across the theological spectrum.

Pastor Samuel Voth Schrag
Peace Mennonite, Dallas

The book challenges the view that "taking climate action *and* maintaining a robust economy with strategic energy independence is impossible" and that "these climate actions would harm economic prosperity and require burdensome personal sacrifice." We will need voluntary action, simplified lifestyles, and laws and regulations enforced at all levels.

Richard Yoder, PhD
Professor of Business and Economics Emeritus,
and International Development Advisor

Dr. Pardini's expertise in both science and theology lends him unique authority. This book is designed to help readers grasp the urgency of taking informed climate action. Furthermore, it emphasizes how our choices today will have a profound impact on the well-being of future generations. The book is a valuable resource for pastors and study groups.

Mark Keller
retired pastor and overseas education-development worker

This is a hopeful book and a great study resource. Steve Pardini brings his scientific background to his study of the climate crisis and its effective solutions. He then draws on his Christian practice and theological studies in a robust discussion of environmental stewardship, spirituality, and faith-based action.

Earl Zimmerman, PhD
environmental activist, religion scholar,
gardener, and retired Mennonite pastor

Based on his unique background in both science and theology, Dr. Steven Pardini presents a unique combination of scientific findings and theological reflection. Through careful and logical exposition, he introduces the basic science of climate change and its impacts on our planet, leading to a theology of climate care and climate justice. The book will serve as a valuable introduction to the science of climate change, a road map for individuals wanting to know what they can do, and a guide for the church to offer a message of hope. The book is suitable for both individual and group study. Guidelines for having constructive conversations, along with questions for reflection and discussion, will provoke thought in readers from diverse backgrounds. This is a valuable resource for individuals, Sunday school classes, and church groups.

Carey K. Johnson
Professor Emeritus of Chemistry, University of Kansas

Steve brings his big heart, formed in the Christian tradition, and his keen mind, trained in scientific practices, to the significant moral, social, and scientific challenges of our time. This is a comprehensive examination of the scientific, political, and economic issues associated with climate change. This could be used as a reference book for many of the facts. However, he does not stop there, and moves us into the spiritual and moral component, filling the book with scriptural support for creation care while also encouraging us in how to dialogue about this complex and challenging problem.

Douglas Day Kaufman
Mennonite Pastor
environmental activist
and executive director of Anabaptist Climate Collaborative

Climate Change and the Healing of Creation

Scientific and Theological Foundations for Creation Care

Steven P. Pardini, PhD, MDiv

Climate Change and the Healing of Creation
Scientific and Theological Foundations for Creation Care

International Standard Book Number: 979-8-9992295-0-2

"We are the first generation to feel the effects of climate change and the last generation who can do something about it."

President Obama

TABLE OF CONTENTS

ii

PREFACE

This book combines rational science with spiritual faith to inform and inspire readers, aiming to transform their minds and hearts. It encourages readers to seek the best knowledge, critically reflect, and take informed action. Additionally, this book carries a message of hope and includes a call to action.

I hope readers find their direction and voices, sharing words of hope and inspiration with those needing transformation. Let us walk alongside one another and collaborate to participate in God's plan to heal all creation. Together, we can bring real improvements to our lives and future generations.

ACKNOWLEDGMENTS

I want to thank Professor Andrew Suderman for his guidance in writing the original paper that inspired this book. As part of the final project in his Liberation Theology class at Eastern Mennonite University, Dr. Suderman encouraged me to submit a paper on climate justice for the Global Mennonite Peacebuilding Conference and Festival III. This paper led to leading chapel services, preaching sermons, doing workshops, teaching classes, and writing this book.

My work led me to connect with the Anabaptist Climate Collaborative, which led me to participate in a yearlong Anabaptist Pastoral Care for the Climate Cohort. This experience inspired me to significantly expand and refine the text of this book. I received valuable feedback from the cohort members, which deepened the content and introduced new ideas that broadened the message.

I then used the book as a thirteen-week curriculum for the Sinapi Discipleship Community at Harrisonburg Mennonite Church. This group helped refine the material to ensure it would effectively serve as a conversation and study guide for discussion groups.

I am deeply thankful for the numerous ways that many friends have contributed to making this book a reality.

I am thankful for my wife, who formatted the manuscript for publication, and for my editor, Russ Eanes, who assisted in publishing.

INTRODUCTION TO

CLIMATE CHANGE AND THE HEALING OF CREATION

This book is a vital study resource designed to inform and guide readers in understanding the urgent need for climate action and the healing of creation. It presents the stark realities of human-caused climate change and the spiritual and moral implications of destroying life-giving ecosystems. The urgency of this situation cannot be overstated. The Earth has been warming rapidly for the last century; 2024 was the hottest year on record. Unless immediate global action is taken, temperatures will continue to rise, and we will face irreversible consequences. However, there is hope. We must address the reality of global warming *today* for our children and grandchildren, who must bear the consequences of our actions. Their lives and livelihoods depend on our immediate and meaningful climate action. It is our responsibility to act now. We can make a difference and bring about the change we need.

Based on climate change policies currently being implemented, climate models predict up to 4.9°C temperature rise above pre-industrial levels by the end of the century.[1] This level of global warming could lead to irreversible climate tipping points, such as the catastrophic melting of Greenland and Antarctica's land-based ice sheets, the thawing and decomposition of boreal permafrost, the decline of the Amazon rainforest, the extinction of coral reefs, and the disruption of the Atlantic Gulf Stream ocean currents. These events would have a devastating impact, affecting hundreds of millions worldwide.[2] The potential for irreversible consequences is real, and we must act now to prevent them.

In the last 200 years, the level of CO_2 in the Earth's atmosphere has increased from about 280 to 420 parts per million (ppm) due to the burning of fossil fuels. This increase has directly contributed to a 1.1°C rise in global temperatures and about a one-foot rise in sea levels. If we surpass the tipping point of 1.5°C, we may trigger self-reinforcing feedback cycles that would accelerate climate change.

If we continue business as usual, the global temperature could increase by an estimated 4.7°C by the end of the century, and the sea level could rise by as much as 7 feet. This significant rise in sea level could impact people living within 60 miles of the coast, including tens of millions in the U.S. and hundreds

---

[1] Climate Action Tracker (2022). The CAT Thermometer. November 2022. Copyright © 2022 by Climate Analytics and New Climate Institute. All rights reserved.

[2] Armstrong McKay, David I., et al., "Exceeding 1.5 C Global Warming Could Trigger Multiple Climate Tipping Points," *Science*, 377 Issue 6611, (September 9, 2022), 1171.

of millions globally, about 40% of the human population. Severe flooding in coastal regions would damage homes, industry, and infrastructure, leading to mass migration and potential conflict.[3]

This book comprehensively analyzes the causes and effects of the climate crisis. It delves into the challenges humanity faces in addressing the current rate of climate change. It critiques the moral frameworks of economic, political, and religious systems that contribute to excessive consumption and the marginalization of those living in poverty due to the exploitation of natural resources. The book also presents a theological foundation for addressing the four-fold alienation of creation and proposes ways to facilitate healing. It discusses the urgent warnings about the destruction of vital ecosystems, offering doctrinal insights and practical approaches to remedying the harm caused by human actions. Additionally, it presents a perspective on environmental stewardship, ecojustice, and ecospirituality. The book's overarching message is one of hope, advocating for individual and collective action to drive sustainable and equitable environmental practices. It calls for initiatives to reduce carbon footprints, urging immediate action to support developing and adopting effective solutions, fostering new personal habits, and advocating for climate-responsible government policies and business operations.

---

[3] Larter, Robert, "Doomsday Glacier 'Holding on by Its Fingernails' – Spine-Chilling Retreat Could Raise Sea Levels by 10 Feet," *SciTechDaily*, University of South Florida, (September 7, 2022)., accessed September 15, 2022.

ABOUT THE AUTHOR

Steve is involved in various activities related to adult education and spiritual growth. He creates curriculum, teaches discipleship classes, leads small groups, youth workshops, and men's Bible studies, and delivers sermons and chapel services at Mennonite churches.

Steve has a PhD in Physical Chemistry and an MDiv. In this work, he uses his knowledge of science and theology to create an informative, inspiring, and compelling study on climate change and climate action.

Part I

The Science of Climate Change

Chapter 1. Cause and Effect of CO2 Emissions

Burning fossil fuels produces carbon dioxide (CO_2), a greenhouse gas (GHG). The buildup of GHGs in the Earth's atmosphere warms the planet. Scientific studies of climate change have correlated human burning of fossil fuels to the release and buildup of CO_2 in the atmosphere. Scientists have determined the effects of global warming on the climate and rising sea levels.

Since the mid-1800s, scientists have known the causes and effects of releasing CO_2 into the atmosphere from burning fossil fuels. Modern scientific measurements have confirmed the early findings. Big data sets have been formulated into sophisticated climate models, making quantitative predictions of the impact of increasing atmospheric CO_2 levels on the climate. Here, we will show how scientists have demonstrated that human burning of fossil fuels is the primary source of climate change and the effects climate change has on the earth's ecosystem.

Earth is Our Only Home

A NASA astronaut aboard the International Space Station took this photo of the earth's horizon, shown in Figure 1. The sun is above the earth, white clouds are over the sea, and the land is in the foreground. This photo of the planet provides a perspective on its beauty—it reminds us that the earth is our only home and that we must care for its life-giving ecosystems.

Figure 1: The Earth's horizon viewed from the International Space Station[4]

---

[4] NASA, "The sun shines above the earth's horizon as the International Space Station orbited 264 miles above the Canadian Provence of Quebec," *Phys.org* (December 21, 2021), accessed April 21, 2022.

The earth's breathable atmosphere, the thin blue layer of gas seen in the picture, is less than 7.5 miles thick (about 40,000 feet—commercial flights in the U.S. typically fly in the 30,000 to 40,000 altitude range). It is the beautiful blue sky we see when we look up in the daytime. This thin layer of gas sustains life on Earth. We use this precious gift of God's creation as a toilet, pumping billions and billions of tons of "waste" carbon dioxide (CO_2) into it every year.

CO_2 is a gas produced when fossil fuels are burned. Humans burn fossil fuels primarily for electricity, transportation, and heat. Pumping excessive CO_2 waste into the atmosphere causes irreversible damage to the earth's life-giving ecosystems. This behavior is unsustainable!

How CO₂ Acts as a Greenhouse Gas

We have known about the CO_2 effect on the atmosphere since the mid-1800s. In 1856, Eunice Foote experimented with air and CO_2 mixtures, as depicted in Figure 2. She took some sealed glass jars and filled them with air. Then, she added ever-increasing amounts of CO2 gas to some jars. She then placed the jars in the sunlight and measured the gas temperature in each jar. She found that the gas temperature increased to a higher level in the jars with CO_2. When she removed the jars from the sunlight, the jars with CO_2 took much longer to cool down.[5] Foote then correlated increased CO_2 concentrations in the earth's atmosphere with increased global temperatures. "An atmosphere of that gas [CO_2] would give to our earth a high temperature; and if, as some suppose, at one period of its history, the air had mixed with it a larger proportion [of CO_2] than at present, an increased temperature from its own action [...would have] necessarily resulted."[6]

---

[5] Foote, Eunice, "Circumstances Affecting the Heat of the Sun's Rays," *The American Journal of Science and Arts*, vol. XXII, (November, 1856): 383. May be found on Google Books.

[6] Foote, Eunice, ibid.

Figure 2: Air + CO₂ Experiments.

"In 1896, the Swedish chemist Svante Arrhenius worried that the increased burning of coal, oil, and firewood was adding millions of tons of carbon dioxide to the atmosphere. 'We are evaporating our coal mines into the air,' he wrote. The result would be a change in transparency of the atmosphere that could heat the planet to intolerable levels."[7]

The experimentally determined absorption spectrum of CO_2 indicates that CO_2 effectively absorbs infrared (IR) radiation, a heat-producing form of light. This means CO_2 acts like a "blanket." The "blanket thickness" is proportional to the amount of CO_2. The greater the CO_2 level, the more heat-producing infrared light is trapped, and the warmer the climate becomes.

The earth absorbs the sun's visible light and emits infrared light. Much of the earth's emitted infrared light escapes harmlessly into space. As the CO_2 level in the atmosphere increases, the amount of infrared light that escapes into space decreases since the CO_2 absorbs it. Thus, CO_2 acts as a greenhouse gas, retaining the heat-producing infrared light in the Earth's atmosphere. The heat of the infrared light held by the CO_2 is then transferred to the land and the oceans, causing their temperatures to increase.

---

[7] Snyder, Howard A., *Salvation Means Creation Healed: The Ecology of Sin and Grace* (Eugene: Cascade Books, 2011), 86.

Figure 3: Earth's Absorption and Emission of Light.

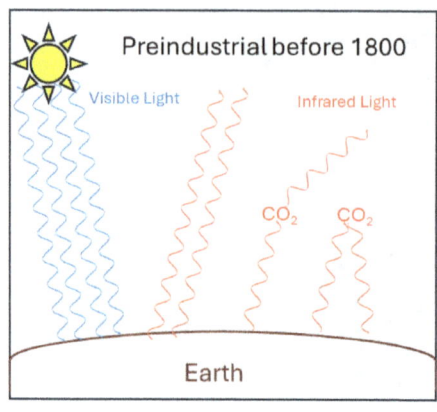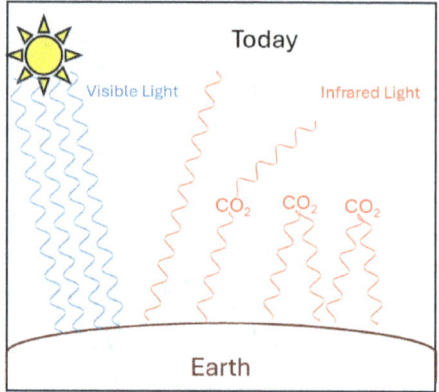

Foote and Arrhenius' predictions have been adequately demonstrated scientifically. Modern scientific measurements and atmospheric modeling have identified human activity as the source of increased CO_2, quantitatively correlated the atmospheric CO_2 level to global temperature changes, and established the impact of this temperature increase on the Earth's climate and ecosystems. The climate models have been improved and provide ominous predictions based on various human behavior scenarios.

The Earth is Rapidly Approaching
the Climate Tipping Point

As CO_2 levels continue to rise, the possibility of irreversible changes to the planet's climate and ecosystems becomes more likely. Climate scientists have identified a 1.5°C (2.7°F) global temperature increase above preindustrial levels as the tipping point where the climate changes become irreversible and may trigger feedback cycles (more on this later) that increase the pace and impact of climate change. Crossing the 1.5°C critical threshold will likely severely impact human civilization.

International climate scientists now accept the following results:

1. Scientists have shown, through atmospheric measurements, that increased levels of CO_2 in the atmosphere increase the amount of heat trapped by the atmosphere. This is known as the greenhouse effect.[8]
2. The atmospheric CO_2 level has increased from pre-industrial levels of 270 ppm in the 1750s to 408 ppm in 2018.[9]
3. The average global temperature has increased by 1.1°C since the 1850s.[10]
4. This climate change has been directly correlated to the burning of fossil fuels.[11]
5. If humans continue burning fossil fuels at current rates, models predict that the average global temperature will increase by about 4.7°C relative to preindustrial times by the end of the century.[12]
6. This temperature rise will have devastating effects on human life: weather extremes, sea-level rise, a decline in food production, increased disease transmission, forced migration, conflict, etc.[13]

The increase in the frequency of severe weather events, record low Antarctic Sea ice, severe droughts coupled with massive forest fires, record ocean temperatures, rapid melting of the Antarctica and Greenland ice sheets, instability in the Atlantic Ocean currents (AMOC: Atlantic Meridional Overturning Currents), increased rate of sea-level rise, and thawing of arctic permafrost have scientists concerned that the earth is rapidly approaching the tipping point.[14]

Climate scientists have strong evidence that urgent action is needed to mitigate the dangerous effects of climate change. These facts have become compelling reasons for limiting CO_2 emissions. There is still time to act, but that time is rapidly running out.

---

[8] Reidmiller, David, et al., *The Climate Report: The National Climate Assessment – Impacts, Risks, and Adaptation in the United States*, (Brooklyn: Melville House, 2018), 30.

[9] Reidmiller, David, et al., 198.

[10] Reidmiller, David, et al., 200.

[11] Reidmiller, David, et al., 200.

[12] Reidmiller, David, et al., 208.

[13] Reidmiller, David, et al., 12-19

[14] Armstrong McKay, David I., et al., 1171.

Historical CO_2 Levels in the Earth's Atmosphere

Scientists have constructed a record of the atmospheric CO_2 history by sampling gas trapped in layers of ice formed yearly in glaciers and continental ice sheets. Figure 1 plots the atmospheric CO_2 concentration in ppm versus time. The results are rather remarkable.

Figure 1: Earth's CO_2 levels from Ice-Core and Atmospheric Samples[15]

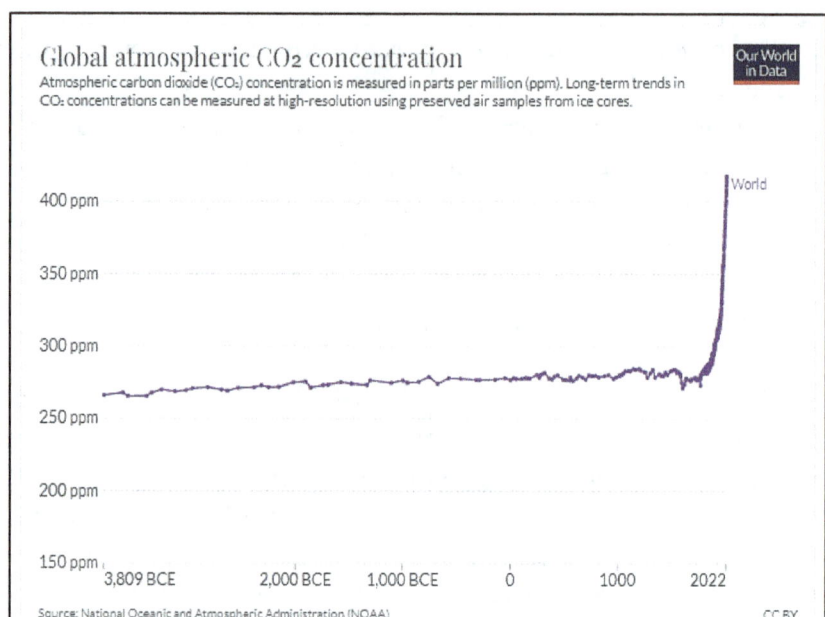

Samples taken from multiple locations around the Earth provide consistent historical data. In addition, twentieth-century atmospheric CO_2 concentrations correlate with modern ice sheet samples. In other words, current atmospheric CO_2 levels correlate with the corresponding year's CO_2 level from the ice sheet. The dramatic upturn in the data depicted in Figure 1 occurred around 1850, marking the beginning of high levels of CO_2 emissions due to the burning of fossil fuels, which provided the energy needed for global industrial expansion.

The data shown in the graph covers about 6,000 years. According to some who analyzed Biblical genealogy data, Adam and Eve were in the garden about 6,000 years ago. At that time, the atmospheric CO_2 level was about 265 ppm.

---

[15] Ritchie, Hannah, Roser, Max, and Rosado, Pablo, "CO_2 Concentrations in the Atmosphere," *Our World Data,* (August 2022), accessed September 15, 2022.

When Jesus walked on the earth some 2,000 years ago, the CO_2 level was about 275 ppm. At the beginning of the Industrial Revolution, a little over 200 years ago, the CO_2 level was about 280 ppm. The atmosphere's CO_2 level hardly changed for thousands of years. However, since the Industrial Revolution began in the 1850s, atmospheric CO2 levels have increased considerably from 280 to 420 ppm. Scientists have shown that this 1.5x increase in CO_2 concentration comes mainly from burning fossil fuels.

Correlating Atmospheric CO_2 Levels with CO_2 Emissions from Burning Fossil Fuels

In Figure 2, the atmospheric CO_2 concentration (blue line) correlates well with the CO_2 emissions from burning fossil fuels (black line). The amount of atmospheric CO_2 has increased proportionally with humans' burning of fossil fuels since the start of industrial expansion in the 1850s. Global CO_2 emissions increased slowly from 1850 to about 5 billion tons annually by 1950. By 2020, the global burning of fossil fuels produced 35 billion tons of CO_2 emissions.

Figure 2: CO2 Emissions and Atmospheric Concentrations (1750-2020) [16]

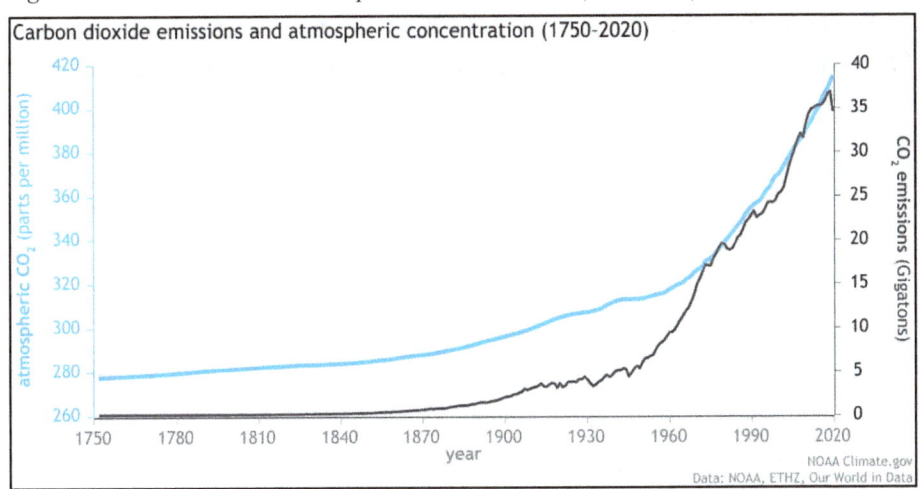

According to physical models and experimental measurements, the recently observed changes in the atmospheric CO_2 level are strongly linked to

---

[16] Diamond, Howard, "Science and Information for a Climate Smart Nation," *NOAA*, (July 8, 2021), accessed December 4, 2021.

human activities such as fossil fuel combustion for energy (including electricity, transportation, construction, and industry) and deforestation (including rainforest burning and land conversion for agribusiness). The current rate of fossil fuel consumption is causing atmospheric CO_2 levels to rise at an unprecedented rate. "Many independent lines of evidence support the finding that human activities are the dominant cause of recent climate change (since 1950)."[17]

Impact of CO_2 Levels on Earth's Temperature

Scientific global temperature measurements correlate well with the atmospheric CO_2 concentration. Over the period shown in Figure 3, 1840-2020, a 1.1°C temperature increase (blue line) has been measured. In addition, the rate of temperature rise has accelerated since the mid-1900s, corresponding to the increase in CO_2 level (orange line) from the burning of fossil fuels.

The 1.1-degree global temperature rise since the 1850s has caused the loss of land-based ice sheets, an increase in sea level rise, thawing of permafrost, year-over-year annual global temperature records, earlier signs of springtime, less snowfall in the winter, increased severity of weather events, droughts in some regions, flooding in other areas, wildfires, and decreased crop harvest. The frequency of severe weather events has increased economic costs and caused human loss of life.

---

[17] Reidmiller, 198.

Figure 3: Global Temperature and Atmospheric CO₂ Concentration (1880-2013)[18]

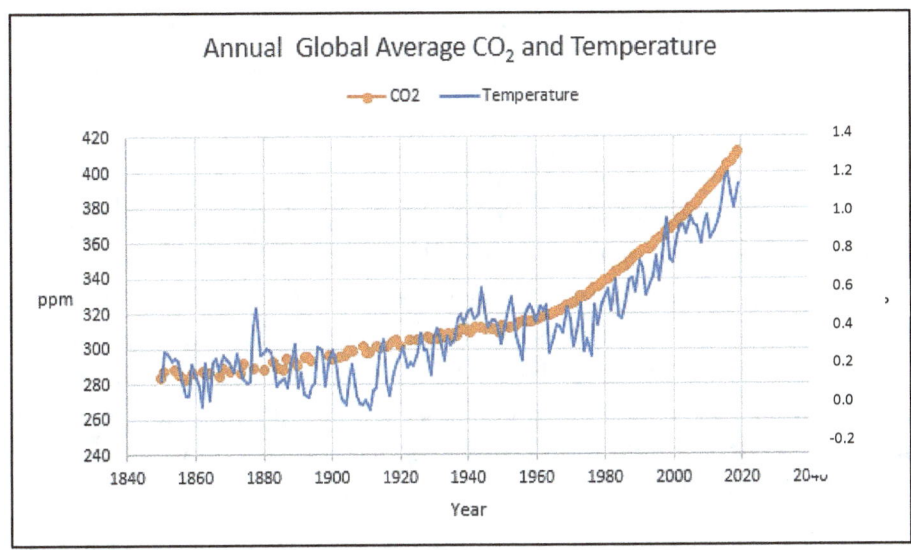

Note: Historical estimates of the Earth's temperatures have been determined from temperature records. Climate scientists have been able to accurately estimate global temperatures using high-altitude tree ring data from around the globe.[19] These measurements correlate well with modern weather station temperature data.

The full effects of the *current* atmospheric CO₂ levels will not be felt for years to come, as it takes time for the heat-producing infrared light absorbed by the CO₂ in the atmosphere to be transferred to the land and oceans. In other words, if we stopped burning fossil fuels today, the land and ocean temperatures would continue to rise for some time and cause significant land-based ice sheet loss.

Due to this lag time, the land and oceans have not yet felt the full impact of the heat already trapped by the CO₂. This lag in response time may give a false sense of plausibility for the denial of or doubt about the cause-and-effect relationship of climate change severity with anthropogenic CO₂ emissions.[20]

[18] Ritchie, Hannah, Roser, Max, and Rosado, Pablo.

[19] Bauer, Bruce, "How Tree Rings Tell Time and Climate History," *NOAA*, (11/29/2018), accessed April 6, 2023).

[20] de Menocal, Peter, "A Climate Science Refresher," *Columbia Climate School*, (March 1, 2017), accessed December 4, 2021.

Impact of Temperature on Earth's Sea Levels:

The oceans will ultimately absorb about 90% of the heat trapped by atmospheric CO_2. Scientists have found that the oceans are warming faster than expected.[21] In 2023, record ocean temperatures were observed, and a low level of Antarctic Sea ice formed. As the oceans warm, the sea level rises due to thermal expansion and melting land-based ice sheets. Land-based ice is found in mountain glaciers and the Greenland and Antarctic ice sheets. As the climate warms, land-based ice melts, flows into the oceans, and raises the sea level.

According to the Environmental Protection Agency (EPA), today's sea levels have risen 9 to 10 inches above pre-industrial levels due to land-based ice melting. "The absolute sea level has risen at an average rate of 0.06 inches per year from 1880 to 2013 (see Figure 4). Since 1993, however, the average sea level has risen from 0.12 to 0.14 inches per year—roughly twice as fast as the earlier long-term trend."[22]

Figure 4: Global Average Absolute Sea Level Change, 1880-2021[23]

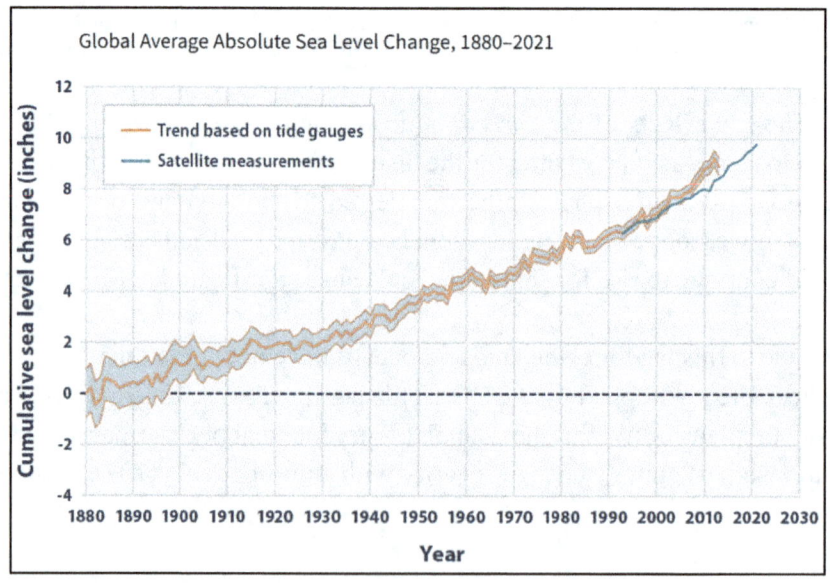

---

[21] Harvey, Chelsea, "Oceans Are Warming Faster Than Predicted," *Scientific American*, (January 11, 2019), accessed September 15, 2022.

[22] Sweet, W. V., et al., "Climate Change Indicators: Sea Level," *United States Environmental Protection Agency*, (updated July 2022), accessed September 15, 2022.

[23] Sweet, W. V., et al.

These findings were determined from satellite data and historical tide gauge data. Corrections were made to account for the sinking of the ocean floor.

According to the National Ocean Service, "About 2 feet of sea level rise along the U.S. coastline is increasingly likely between 2020 and 2100 because of emissions to date. Failing to curb future emissions could cause an additional 1.5 to 5 feet of rise for a total of 3.5 to 7 feet by the end of this century."[24]

"According to the United Nations, roughly 40% of the human population lives within 60 miles of the coast."[25] The projected magnitude of the rising sea level would cause severe flooding in coastal regions and waterways up to 60 miles inland, damaging homes, industry, and infrastructure. It would force mass migration for "tens of millions of people in the U.S. and hundreds of millions globally."[26]

Rising sea levels are a "canary in the coal mine." Scientists are seeing the leading edge of an "irreversible change." Coastal residents have experienced the effects of climate-caused sea level rise: flooding, high tides, storm surges, forced migration, and property loss.

The loss of land-based ice sheets, an irreversible climate tipping point, will lead to severe and catastrophic impacts. Reduction in CO_2 emissions is essential to averting a climate catastrophe.

Importance of Protecting the Antarctic and Greenland Ice Sheets

One feature of global temperature rise is that the temperature at the poles increases several times faster than the global average, exposing the ice sheets in Greenland and Antarctica to a greater temperature rise. Enough land-based ice is stored in the Greenland and Antarctic ice sheets to raise the ocean level by up to 10 feet. A sea level rise of this magnitude will affect major coastal metropolitan areas, causing flooding due to high tides, heavy rainfall, and storm

---

[24] Sweet, W.V., et al., "Global and Regional Sea Level Rise Scenarios for the United States," *National Oceanic and Atmospheric Administration*, (February 2022), accessed October 20, 2022.

[25] Larter, Robert, "Doomsday Glacier 'Holding on by Its Fingernails' – Spine-Chilling Retreat Could Raise Sea Levels by 10 Feet," *SciTechDaily*, University of South Florida, (September 7, 2022)., accessed September 15, 2022.

[26] Sweet, W.V., et al., "Global and Regional Sea Level Rise Scenarios for the United States," *National Oceanic and Atmospheric Administration*, (February 2022), accessed October 20, 2022, p. 15.

surges, as shown in the "Sea Level Rise viewer." [27] In the U.S. alone, "researchers project that 28,800 square miles of land would be affected where 12.3 million people live; [in some] 40 large U.S. cities more than half of the land area falls less than 10 feet above the high tide line." [28] New Orleans, Miami, Boston, New York, Norfolk, Virginia Beach, Charleston, and many other coastal cities in the U.S. will be the most affected.[29]

Recent scientific studies have revealed that the Thwaites glacier (known as the "doomsday glacier") in West Antarctica is melting faster than predicted due to warmer ocean temperatures.[30] "The glacier is holding on today by its fingernails, and we should expect big changes over small [geological] timescales in the future."[31] The melting of the Thwaites glacier, roughly the size of Florida, will significantly impact global sea levels. The Thwaites Glacier is a wide glacier that acts as a keystone at the leading edge of the West Antarctic ice sheet. If Thwaites collapses, the rest of the West Antarctic ice sheet will flow into the ocean. According to geoscientist Alastair G.C. Graham, the ramifications of melting the Thwaites glacier are considerable: "You can't take away Thwaites and leave the rest of Antarctica intact."[32] It is uncertain how fast the ice sheet may move into the ocean. The latest scientific studies indicate that the ice is less stable and moving faster than expected. According to Graham, reducing greenhouse gas emissions is critical to lowering the risk of melting the West Antarctic ice sheet. "Right now, we can do something about it— especially if we can stop the ocean from warming. The potential impact of the loss of the Thwaites glacier and surrounding ice could raise sea level from 3 to 10 feet."[33]

---

[27] NOAA, "Sea Level Rise Viewer," *National Oceanic and Atmospheric Administration*, (updated June 2023), accessed October 14, 2023.

[28] Heltzel, Paul, "Global Warming Right Before Your Eyes," *Seeker*, (Published 5/14/2014) accessed October14, 2023.

[29] Strauss, Benjamin, "What Does the Earth Look Like with 10 Feet of Sea Level Rise?" *Climate Central*, (May 13, 2014), accessed October 14, 2023.

[30] A.G.C. Graham, et al, "Rapid retreat of Thwaites Glacier in the pre-satellite era," *Nature Geoscience* 15, (September 5, 2022): 706–713. accessed September 15, 2022.

[31] Larter, Robert.

[32] Alder, Ben, "'Doomsday glacier' is melting faster than thought, study finds," *Yahoo!News*, (September 7, 2022). accessed September 15, 2022.

[33] Larter, Robert.

Updated models of the melting rate of the Greenland ice sheet indicate rates of "ice flow discharge" greater than earlier predictions. Greenland ice loss due to human-driven climate change significantly contributes to rising global sea levels.[34] Researchers estimate that the "inevitable melting of 3.3% of the Greenland ice sheet" will "trigger nearly a foot of global sea level rise."[35]

These findings indicate the irreversible impact of existing human CO_2 emissions levels. There is still time to act. Reducing CO_2 emissions now is needed to prevent a more significant, potentially catastrophic loss of the Antarctic and Greenland ice sheets.

Impact of Ice Sheet Melting on the Gulf Stream

Natural currents caused by tides, wind, temperature gradients, and salinity differences circulate ocean water around the globe in a large loop. The Atlantic Ocean current, known as the AMOC, shown in Figure 5, keeps the warm Gulf Stream water moving in a loop from the North to the South Atlantic. As the Greenland ice sheet melts, the AMOC circulation weakens. The freshwater from the Greenland ice sheets mixes into the North Atlantic water, lowers the ocean water's saltiness, and slows the AMOC flow. Disruption of the AMOC flow will considerably shift European weather patterns, causing droughts, lower temperatures, and a dramatic fall in crop production.

---

[34] Box., J. E., Hubbard, A., Bahr, D.B., et al. "Greenland ice sheet climate disequilibrium and committed sea-level rise," Nature Climate Change 12 (August 29, 2022), 808, 808-813, accessed September 29, 2022.

[35] Mooney, Chris, "Greenland ice sheet set to raise sea levels by nearly a foot, study finds," *The Washington Post: Climate and Environment*, (August 29, 2022), accessed September 29, 2022.

Figure 5: The Gulf Stream: Atlantic Meridional Overturning Circulation (AMOC)[36]

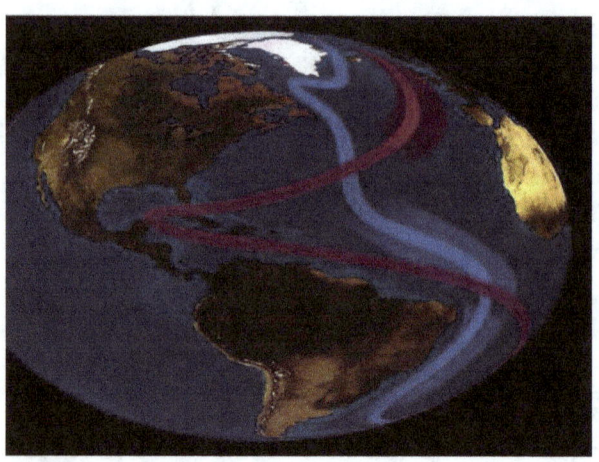

Measurements have indicated that the melting and weakening of the ice sheets have increased alarmingly since the mid-20th century, which has weakened the AMOC. The complete collapse of the AMOC would be one of the irreversible tipping points of climate change.

Although the melting of the ice sheets, catastrophic rise in ocean levels, and interruption of the AMOC are not likely to occur in the 21st century, the CO_2 emissions that will cause these effects are being emitted now. It takes time to transfer the heat from the atmosphere to the land, ice sheets, and oceans. These early warning signs tell us there is time to act, but that time is running out quickly.

There is More than One Type of Greenhouse Gas

Thus far, this book has discussed CO_2 sourced from burning fossil fuels. As mentioned earlier, CO_2 is a greenhouse gas (GHG) because it retains heat-producing infrared light in the atmosphere. However, CO_2 is not the only GHG that contributes to climate change. Another significant GHG produced by human activity (e.g., fossil fuel production and agricultural sources) is methane (CH_4). CH_4, like CO_2, acts as a GHG because it absorbs infrared light. CO_2 and

---

[36] National Ocean Service, "What is the Atlantic Meridional Overturning Current?" *National Oceanic and Atmospheric Administration*, (updated 1/20/2023), accessed October 14, 2023).

CH_4 combined are the most significant physical drivers of climate change.[37] In one sense, that is good news; since human activity produces these gases, changes in human activity can reduce them, thus mitigating their effects on the climate.

Unfortunately, that is not the whole story. The Earth's permafrost holds a potentially significant source of CH_4. Permafrost is "ancient" biological material from plants and animals beneath an ice layer in a "permanently" frozen condition. As the Earth warms, ice melts, permafrost thaws, the biological material decomposes, and CH_4 is released. CH_4 released from decomposing permafrost further increases the amount of GHG in the Earth's atmosphere beyond the direct human activity sources of fossil fuel production and agriculture. Future CH_4 release from permafrost will amplify the rate of human-induced GHG emissions.

Much of the Earth's permafrost is located in the Arctic region. Scientists have shown that the rate of warming is greater in the Arctic region, and significant amounts of CH_4 are now being released from the permafrost.[38] Permafrost CH_4 emissions currently "act like a headwind [that our climate change] mitigation efforts are fighting against."[39]

Table 1 compares the concentration change of CO_2 and CH_4 in the Earth's atmosphere during the Industrial Revolution from 1850 to today. Human activity (production and burning of fossil fuels and agriculture) during the industrial age has increased CO_2 and CH_4 GHG levels significantly: CO_2 levels have risen from 280 to 420 ppm, a 1.5x increase, and CH_4 has increased from 0.70 to 3.65 ppm, a 5.2x increase. CO_2 remains in the atmosphere for more than 2,000 years, while CH_4 has a 12-year lifetime. Although CO_2 is at a much higher concentration and chemically more stable than CH_4, CH_4 traps 34 times more infrared heat than CO_2, making $CH4$ a potent GHG. Together, CO_2 and CH_4 produce about 90% of human-caused climate change.

---

[37] EPA, "Greenhouse Gas Emissions," Unites States Environmental Protection Agency, (updated April 2023), accessed August 8, 2023.
[38] Hayhoe, Katharine, *Saving us:* 104.
[39] Hayhoe, Katharine, *Saving us:* 105.

Table 1: Comparison of Greenhouse Gases (GHG) CO_2 and CH_4[40]

GHG	Pre-Industrial Level (ppm)	Current Level (ppm)	Average Lifetime (years)	Warming Potential
CO_2	280	420	>2,000	1
CH_4	0.70	3.65	12	34

Note: Later in the book, we will mention refrigerants, powerful GHGs with >1,000x CO_2 warming potential, but at much lower atmospheric levels.

Permafrost melting has dramatically impacted people in the Arctic region. Thawed permafrost becomes unstable ground that no longer supports homes, roadways, and other infrastructure. The decomposition of permafrost significantly increases atmospheric CH_4 levels. Permafrost thawing and decomposition leading to the release of CH_4 is another "canary in the coal mine." The Earth is at the leading edge of a climate tipping point, with irreversible permafrost loss.

Impact of Feedback Cycles on Climate Change

The release of CH_4 from permafrost decomposition can create a climate-change feedback cycle. Ice, a good sunlight reflector, covers the permafrost and protects it from the sun's warming rays. Loss of ice exposes the permafrost directly to sunlight, and the exposed permafrost begins to warm. When permafrost warms, it decomposes and releases CH_4. Since CH_4 is a GHG, its release increases atmospheric warming, increasing the rate of ice melting and exposing more permafrost, which increases the release of CH_4. This is a feedback cycle: loss of ice increases the rate of CH_4 release, and CH_4 release increases the rate of ice loss.

Exceeding the Paris Agreement limit of 1.5°C by the end of the 21st century and the Conference of the Parties COP26 goal of net-zero emissions by 2050 will likely result in the "collapse of the Greenland and West Antarctica ice sheets… and trigger widespread permafrost thaw."[41] Once the 1.5°C tipping point is crossed, feedback cycles may become self-reinforcing, leading to abrupt climate changes such as the "sudden" collapse of the Gulf Stream

---

[40] Wiki, "Greenhouse Gas," Wikipedia, (updated August, 2023), accessed August 17, 2023.
[41] Armstrong McKay, David I., et al., 1171.

circulation system. The Earth is fast approaching the tipping point. Human activity is the source of the problem, and human action is needed to avert disaster for future generations.

Chapter 2. Sources and Uses of Fossil Fuels

Highly accurate data are available to provide a historical quantitative description of the types of fossil fuels used, what they are used for, how much they have been used, and by whom they have been used since the mid-1800s. When entered into modern climate change models, these data may be used to predict the relative impact of fossil fuel consumption globally. These models predict the magnitude of the increase in temperature based on different fossil fuel consumption scenarios from now to the end of the century. Here, we will show how the rate of human fossil fuel consumption over the next 75 years will dramatically impact the global temperature.

Which Fossil Fuels Do Humans Use?

As shown in Figure 6, most CO_2 emissions come from burning coal, oil, and gas. In 1850, 100% of the 200 million tons of CO_2 emissions came from burning coal. In 2021, CO_2 emissions increased to 35 billion tons, of which 40% came from coal, 32% from oil, and 21% from gas. The remaining 7% came from cement production, flaring, and other industrial activities.

Figure 6: CO_2 Emissions by Fuel Type[42]

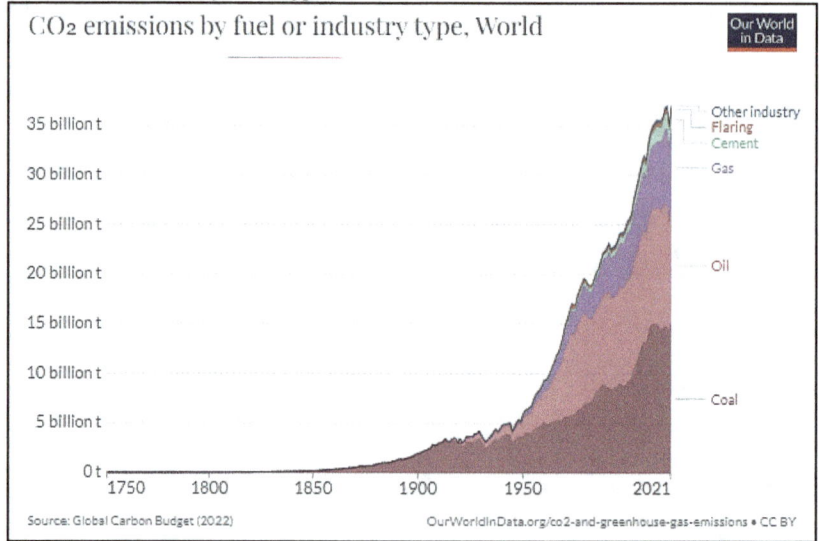

---

[42] Ritchie, Hannah, and Roser, Max, "CO2 Emissions," *Our World Data*, (2022) accessed September 15, 2022.

The increased use of fossil fuels facilitated the global economic expansion after World War II. Oil use grew dramatically from 1950 onwards. Later, natural gas supplemented the demand for fossil fuels. Yet burning coal today produces more CO_2 emissions than other fossil fuels.

What Do Humans Use Fossil Fuels For?

Figure 7 shows the energy use by sector in the United States in 2019. Some 33% of the U.S.'s GHG emissions come from burning fossil fuels to produce, transmit, and distribute electricity and heat. About 31% comes from burning fossil fuels to transport people and goods by land, air, and sea. These two sectors alone account for nearly two-thirds of US GHG emissions. On-site energy generation to heat buildings and cook food contributes 9.4% of GHG emissions, and energy to run manufacturing plants and do construction contributes 7.4% annually. Note: Electricity use in buildings from utilities is excluded from this sector and is covered in the electricity and heat production sector. This chart indicates that land use management and forestation efforts helped to remove 3.9% of GHG emissions from the atmosphere annually.

These data suggest that potentially significant reductions in GHG emissions may be realized by increasing efficiency for power generation, transportation, building insulation, manufacturing, and construction. Conversion from fossil fuels to clean energy sources for power generation and transportation will also significantly reduce GHG emissions. CO_2 removal from the atmosphere by planting forests, reforesting lands burned for cattle grazing, and planting trees in urban areas to provide localized cooling are also substantial contributors to CO_2 reduction.

Figure 7: U.S. GHG Emissions by Sector in 2019[43]

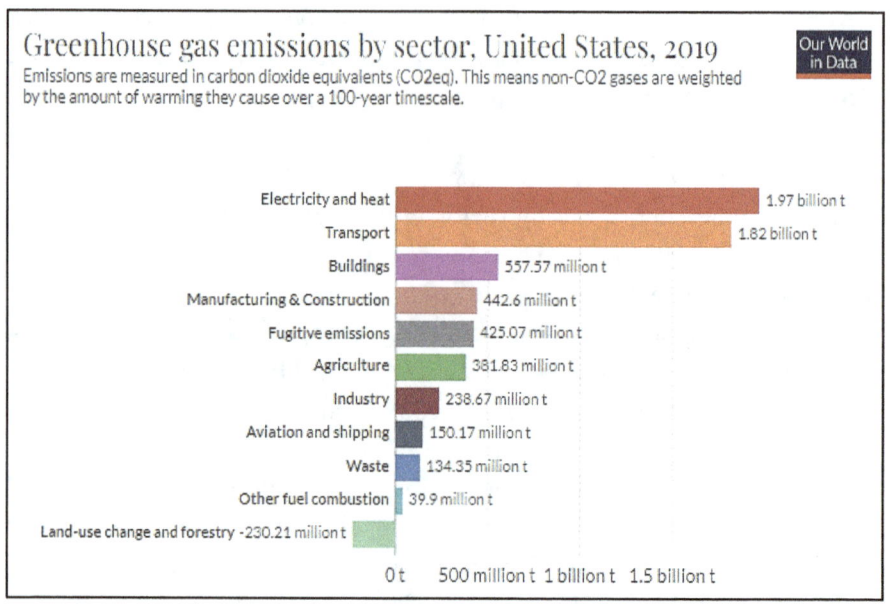

Sector	U. S. 2019 GHG Emissions	
	(billions of tons)	%
Electricity and Heat	1.97	33
Transport	1.82	31
Buildings	0.56	9.4
Manufacturing & Construction	0.44	7.4
Fugitive Emissions (Note 1)	0.43	7.3
Agriculture	0.38	6.4
Industry	0.24	4.0
Aviation and Shipping	0.15	2.4
Waste	0.13	2.5
Other fuel combustion	0.04	0.6
Land use change and forestry	-0.23	-3.9
Total	**6.16**	**100**

(Note 1: from coal mining, oil and gas activity, and refrigerant leakage).

---

[43] Ritchie, Hannah, and Roser, Max.

How Much Total CO₂ Have Humans Produced?

Another important metric is the cumulative CO_2 emissions humans have produced by burning fossil fuels. Figure 8a shows the total CO_2 emissions by country since the start of the Industrial Revolution. These data indicate that Europe and the U.S. have emitted the greatest amounts of CO_2.

Figure 8a: Total (Cumulative) CO₂ Emissions since 1750[44]

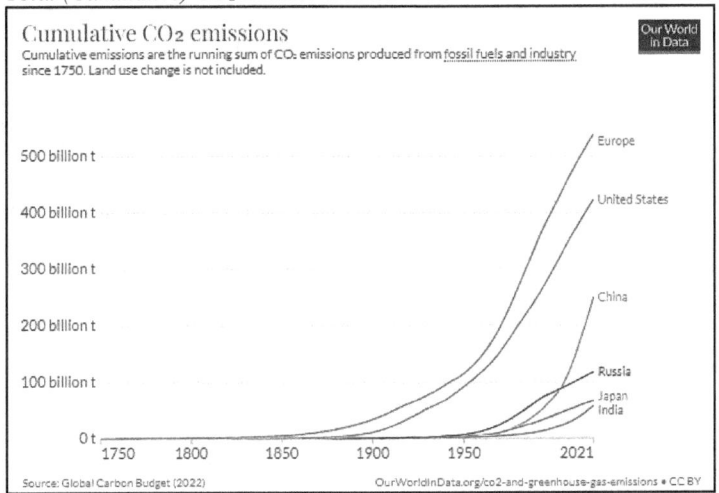

We need to get to net-zero carbon emissions. These lines cannot continue rising at their current pace! A transition to clean energy is required to reduce CO_2 emissions and keep the earth from moving past the 1.5°C climate change tipping point. The nations leading in the clean energy transition will significantly impact climate change and reap substantial economic benefits.

Many countries claim substantial progress has been made. Many have pledged additional climate action but have yet to make plans to implement these changes. Citizens must hold their government to these promises. CO_2 emission reduction programs must not just be talked about; they must be implemented. Time is running short.

Developed countries have commercially available clean energy alternatives that are economically viable and lead to annual CO_2 emissions reductions. As shown in Figure 8b, these technologies are brought online, GHG emissions decline. Developing countries can skip the development of fossil fuel technologies and invest in clean energy technologies. Developed countries can

---

[44] Ritchie, Hannah, and Roser, Max.

24

facilitate the transition from fossil fuels in developing countries by making clean energy technologies commercially available and affordable.

As shown in Figure 8c, per capita emissions in the U.S. remain among the highest in the world but have decreased since about 2000. Note: The U.S. ranks 11th in per capita emissions; many Middle Eastern oil-producing countries are higher. Meanwhile, developing economies such as India have relatively low per capita emissions but are on an increasing trend.

Figure 8b: Annual CO_2 Emissions since 1750[45]

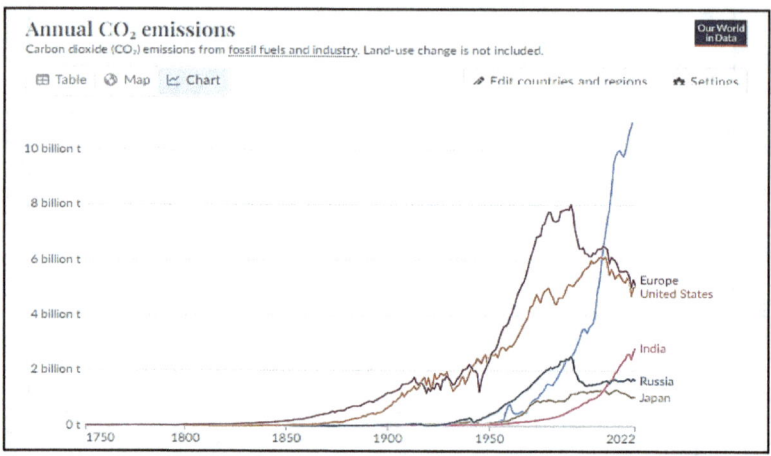

Figure 8c: Per Capita CO_2 Emissions since 1750[46]

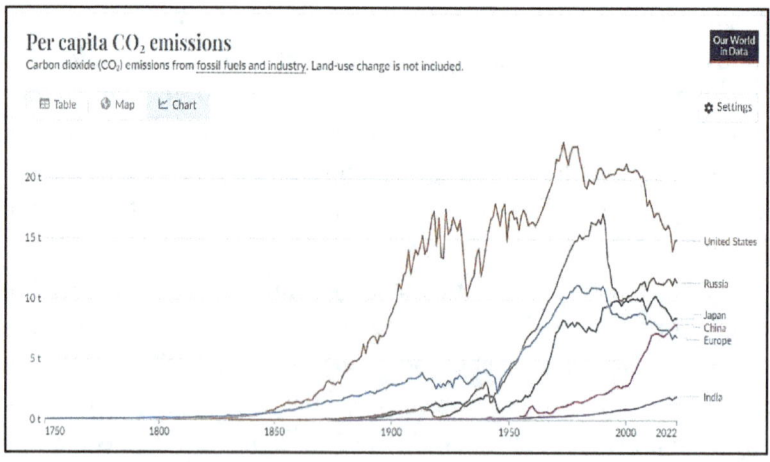

---

[45] Ritchie, Hannah, and Roser, Max.
[46] Ritchie, Hannah, and Roser, Max.

Fossil Fuel Consumption by Nation

Figures 9a - 9f plot the relative CO_2 emissions from burning coal, oil, and gas by source country.[47]

Figure 9a shows the United States' CO_2 emissions since 1850 by fossil fuel type. In 1850, roughly 100% of CO_2 emissions in the U.S. came from burning coal. In 2021, 20% of CO_2 emissions came from coal, 44% from oil, and 33% from burning gas. The peak CO_2 emissions for the U.S. came in 2006 at 6 billion tons. CO_2 emissions have declined as wind, solar, and other renewable energy sources have come into service, and energy efficiency improvements have been implemented.

Figure 9b shows that Europe's CO_2 emissions peaked in 1990 at 8 billion tons, 37% of which came from coal. By 2021, coal emissions decreased to 23%.

Figure 9c shows that Japan's CO_2 emissions peaked in 2012 at 1.3 billion tons, 34% of which came from coal. By 2021, 39% of Japan's emissions came from coal. This increase in coal use was necessitated by the shutdown of the Fukushima nuclear reactor in 2011 due to the Tohoku earthquake and tsunami.

Figure 9d shows that Russia's CO_2 emissions peaked at 2.5 billion tons in 1990, 30% of which came from coal. The falloff in Russia's CO_2 emissions in 1990 came with the breakup of the Soviet Union. In 2021, 22% of Russia's emissions came from coal.

Figure 9e shows China's CO_2 emissions by source. In 2021, China's CO_2 emissions reached 11.6 billion tons, 70% of which came from burning coal.

Figure 9f shows that India emitted 2.7 billion tons of CO_2 in 2021, 67% of which came from coal.

Developing countries have yet to reach their peak CO_2 emission levels and rely heavily on coal. Globally, governments, businesses, and consumers must comply with COP26 goals to "phase down" coal burning and reduce overall CO_2 emissions. Developed and developing countries need to convert to renewable, clean energy sources.

Some countries have made progress in reducing total CO_2 emissions and reducing coal as an energy source. However, as shown in Table 2, burning coal remains the most significant contributor to global CO_2 emissions.

---

[47] Ritchie, Hannah, and Roser, Max.

Table 2: Summary of CO₂ Emissions by Fuel Source[48]

Country	Peak CO₂ Emissions (Billions of Tons)	Year of Peak CO₂ Emissions	2021 CO₂ Emissions (Billions of Tons)	2021 Coal (%)	2021 Oil (%)	2021 Gas (%)
U.S.	6.0	2006	5.0	20	45	33
Europe	8.0	1990	5.0	23	36	37
Japan	1.3	2012	1.1	39	37	21
Russia	2.5	1990	1.8	22	23	50
China	11.5	2021	11.5	69	15	7
India	2.7	2021	2.7	67	23	5
World			37	40	31	21

Europe and the U.S., the greatest historical total CO₂ emitters, have made progress in CO₂ emission reduction, but not at a pace that will meet the Paris goal of 1.5°C by 2050. China's and India's economies are growing, as is their dependency on coal. *All* nations must accelerate the transition from fossil fuel technology and switch to wind, solar, hydro, nuclear, and other clean energy sources to meet their country's energy needs.

Figure 9a: U.S. CO₂ Emissions by Source [49]

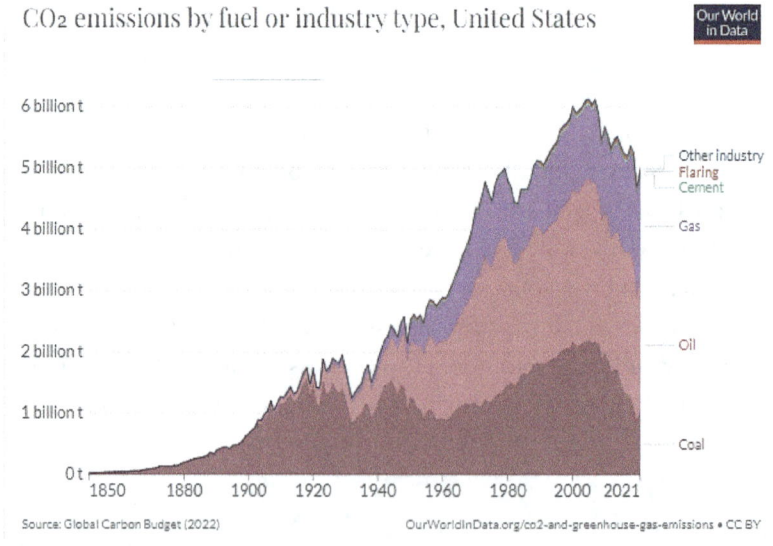

CO₂ emissions by fuel or industry type, United States

Source: Global Carbon Budget (2022)

OurWorldInData.org/co2-and-greenhouse-gas-emissions • CC BY

---

[48] Ritchie, Hannah, and Roser, Max.

Figure 9b: European CO₂ Emissions by Source [49]

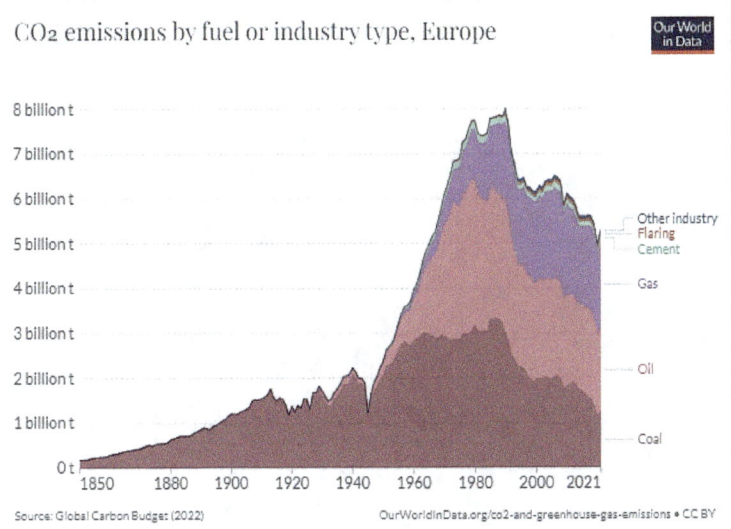

Figure 9c: Japan CO₂ Emissions by Source [49]

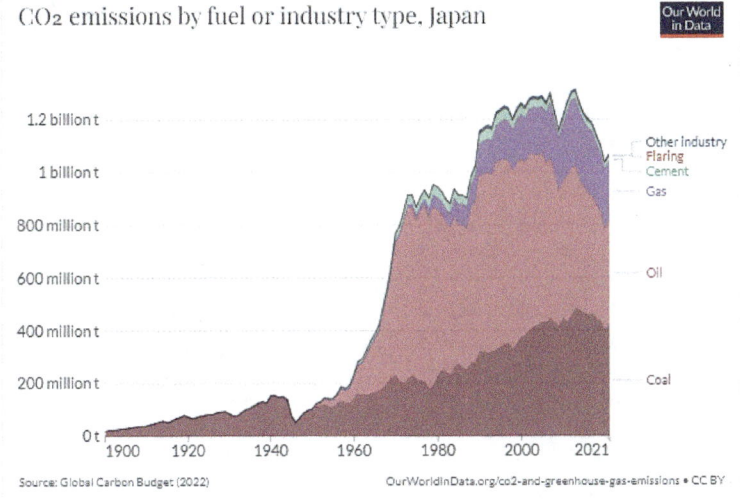

Figure 9d: Russia CO₂ Emissions by Source [49]

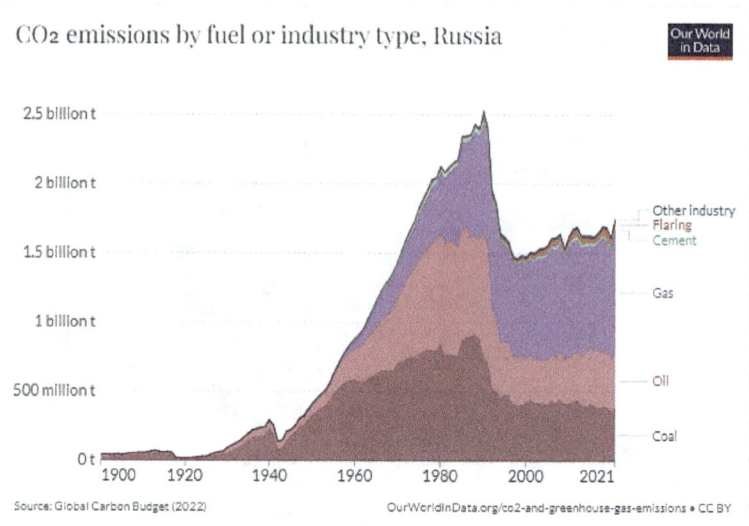

Figure 9e: China CO₂ Emissions by Source [49]

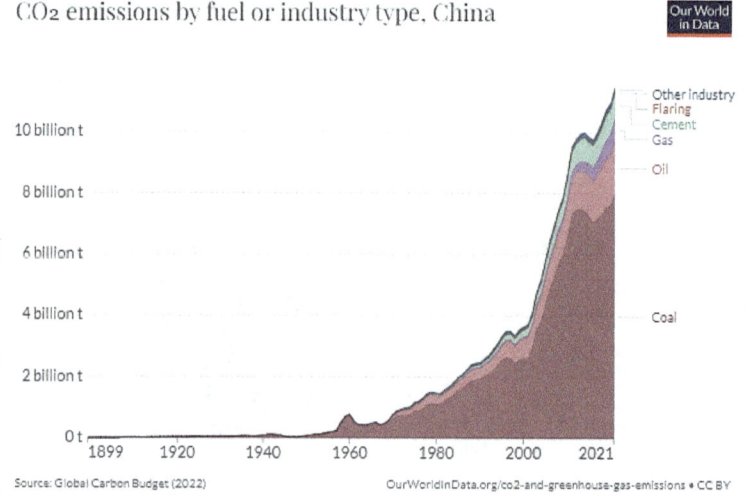

Figure 9f: India CO$_2$ Emissions by Source[49]

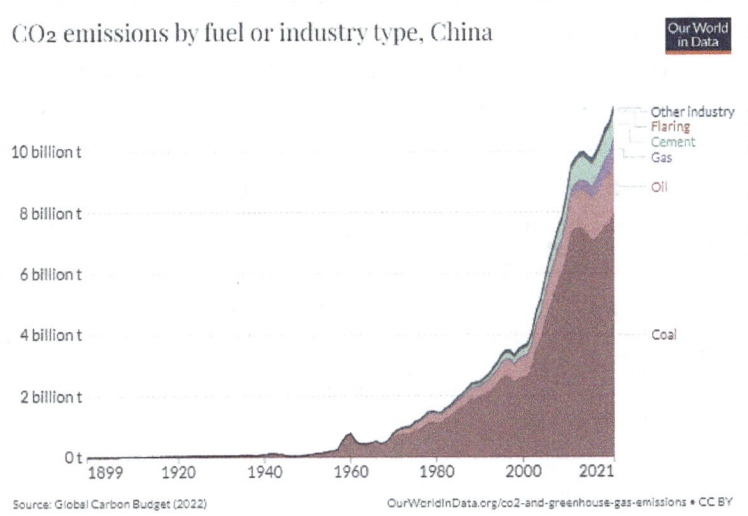

CO$_2$ emissions by fuel or industry type, China

Our World in Data

10 billion t

8 billion t

6 billion t

4 billion t

2 billion t

0 t

1899   1920   1940   1960   1980   2000   2021

Other Industry
Flaring
Cement
Gas
Oil
Coal

Source: Global Carbon Budget (2022)

OurWorldInData.org/co2-and-greenhouse-gas-emissions • CC BY

National Wealth and CO$_2$ Emissions

Figure 10 plots the relationship between national wealth (Gross Domestic Product – GDP) and total carbon emissions (billions of tons per year) by country. On the plot, further to the right indicates greater wealth, and higher up indicates greater CO$_2$ emissions. The size of the circles represents the CO$_2$ emission per capita. The plot compares data from 1990 and 2020. As the economies have grown, carbon emissions have climbed in developing countries. From 1990 to 2020, China's annual per capita GDP grew from $1,424 to $16,316, and yearly carbon emissions increased from 2.48 to 10.67 billion metric tons. China's annual per capita emissions have risen from 2.1 to 7.4 metric tons per year per person. China's economic expansion has relied heavily on coal.

In 2020, the U.S. consumer economy had the highest annual per capita GDP at $59,920 and CO$_2$ emissions at 14.2 metric tons per person. The U.S. has taken essential steps in combating climate change. It has implemented regulations and shifted power generation from coal to cleaner options. From 1990 to 2020, in the US, annual emissions decreased from 5.11 to 4.71 billion metric tons per year, annual per capita emissions fell from 20.8 to 14.4 metric tons, and the economy grew by 48%.

The European Union's mix of consumer, business, and government actions has paid off. Since 1990, the economy has grown 55%, annual per capita emissions have decreased from 9.2 to 5.8 metric tons per person, and annual emissions have dropped from 3.86 to 2.60 metric tons. Europe's efforts indicate that stakeholder coordination produced economic growth and CO_2 emission reductions.

The EU and US data provide examples of decoupling economic growth from CO_2 emissions reductions.

Following the 2011 Fukushima nuclear reactor accident, Japan's consumption of fossil fuels increased. Japan is working its way back to historical per capita CO_2 emission levels. Industry in Russia declined in the 1990s following the collapse of the Soviet Union; the per capita emissions in 2020 are below 1990 levels.

Figure 10: CO₂ Gas Emissions and GDP by Country 1990 versus 2020[49]

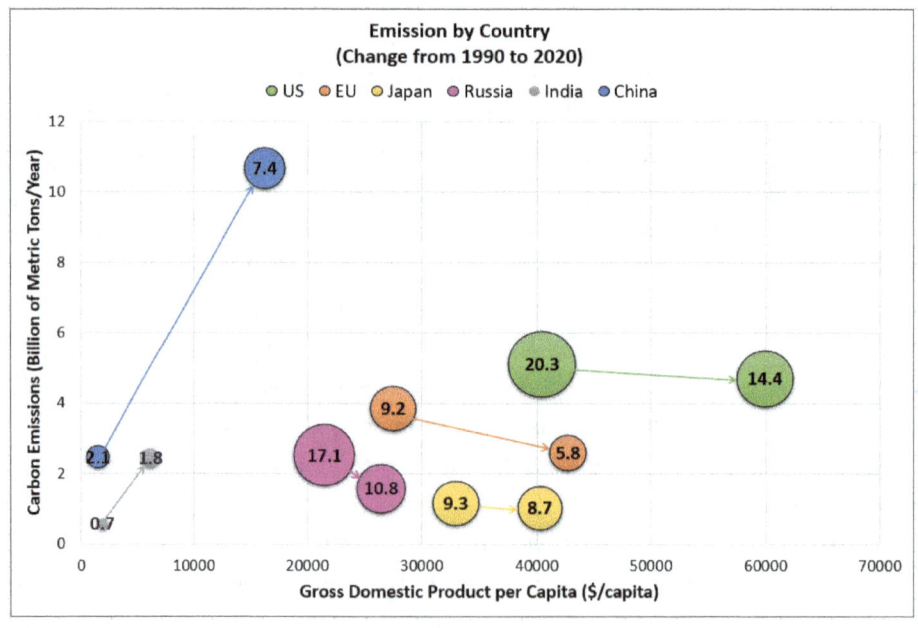

[49] Ritchie, Hannah, Roser, Max, and Rosado, Pablo, "CO₂ Emissions," *Our World Data,* (April 2022), accessed September 15, 2022.

Climate Change Scenarios

Figure 11 plots the impact of climate action programs on predicted greenhouse gas emissions and global temperature changes. In 2020, GHG emissions (CO_2, methane, and other GHGs) totaled 50 billion tons/yr. In Figure 11, the 1.5°C Paris Agreement target requires net-zero GHG emissions by 2070, which includes carbon neutrality. Carbon neutrality means the amount of CO_2 emissions equals the amount of CO_2 removed from the atmosphere each year. To achieve carbon neutrality, CO_2 emissions must be reduced by converting fossil fuels to wind and solar energy sources, and CO_2 must be removed from the atmosphere through land use, forestry, sustainable agriculture practices, and capturing CO_2 in smokestacks. Actions taken *now* by consumers, businesses, and governments will determine our climate future.

The average global temperature in 2020 is 1.1 °C above preindustrial levels. If no climate policies are enacted, the worst-case scenario predicts a temperature increase of 4.1 to 4.8 °C by 2100. If current climate policies are enacted, such as increasing manufacturing efficiency, power generation energy efficiency, building insulation, and transportation fuel efficiency, climate models predict a 2.5 to 2.9 °C temperature rise. Implementation of current climate actions and future pledges, such as converting energy production from burning fossil fuels to clean energy (e.g., wind and solar) and replacing fossil fuels for transportation (e.g., by using clean energy to charge vehicle batteries), results in an estimated 2.1 °C global temperature increase. Additional climate actions are needed to reach carbon neutrality by 2050, such as using carbon capture and planting trees to remove CO_2 from the atmosphere. These are expected to limit the predicted global temperature rise to 1.5 to 2°C above preindustrial levels by 2100.

Figure 11: Global Greenhouse Gas Emissions and Warming Scenarios[50]

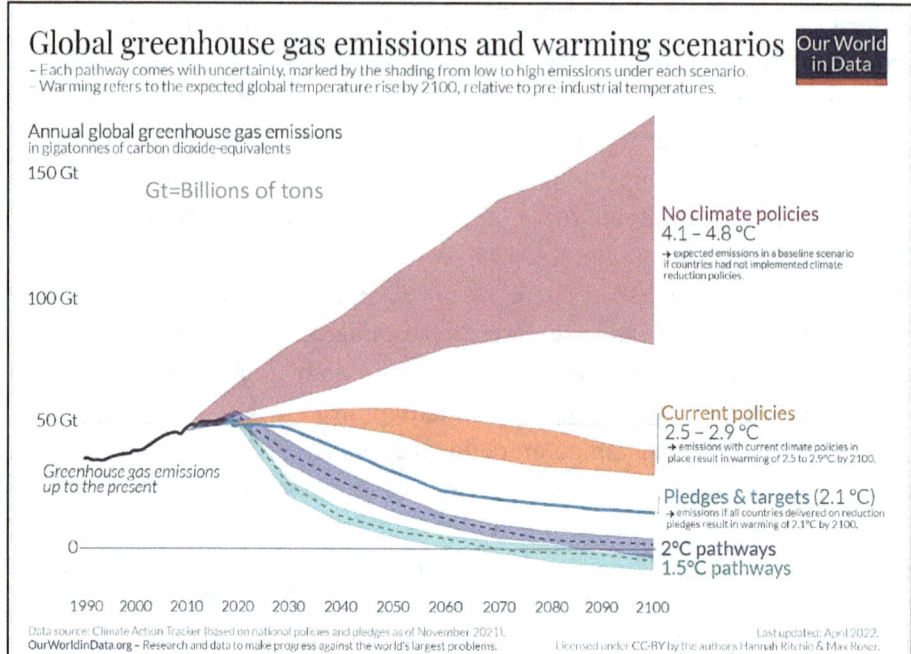

The 2016 Paris Agreement Goal is to limit the global temperature increase to 1.5°C above pre-industrial levels.

COP26's (Congress of the Parties) goal requires carbon neutrality by 2050 and net-zero greenhouse gas emissions by 2070.

All of the actions described above must be implemented. The Intergovernmental Panel on Climate Change (IPCC) states that the likelihood of "temperature increase in this century can be limited to 2°C, increases significantly when greenhouse gas emissions are curbed sharply in the near term."[51] This is a responsibility we all share.

---

[50] Ritchie, Hannah, and Roser, Max, "Future Greenhouse Gas Emissions Scenarios," *Our World Data,* (April 2022), accessed September 15, 2022.

[51] Lelieveld, J., et. al, "Effects of fossil fuel and total anthropogenic emission removal on public health and climate," *Proceedings in the National Academy of Sciences,* April 9, 2019, 116 (15): 7192.

Part II

The Struggle to Make Progress

on Climate Change

Chapter 3. Transition to a Clean Energy Economy

Our shift to clean energy is not just a far-off vision but an achievable reality. The technologies available can offer renewable, clean, and sustainable energy sources that can effectively power our electric grid and transportation systems. Economic cost analysis and energy efficiency calculations confirm the practicality of this model for transitioning to clean energy. We have the necessary tools and expertise to accomplish this transition.

The energy that comes from a naturally replenished source is renewable. Clean energy comes from sources that emit little or no greenhouse gases. Energy sources are sustainable if the supply is sufficient to meet the needs.

Renewable, clean, sustainable energy sources, such as wind, sunlight, movement of water, and geothermal, supply energy to a wide range of power-producing technologies, including wind farms, solar panels (photovoltaics, PV), hydroelectric dams, geothermal, solar-thermal (solar concentrating), and ocean waves. Biomass (organic material from plants and animals) is a renewable energy source, but not sustainable since biomass has a low fuel value and its use exceeds its production rate. Nuclear energy is not renewable since it depends on a finite supply of uranium, but it is a low-CO_2-emitting source.

Here, we will show that the global transition to clean energy is a matter of choice and a pressing need. The pace of this transition has been alarmingly slow, and it is time for humans to step up. We have the knowledge, technology, and capabilities to accelerate this transition significantly. What we lack is the willingness to do it. If we want to avert the worst consequences of climate change, we must act now. To meet the climate goals of 1.5 to 2°C above preindustrial levels by 2100, we must speed up clean energy development to electrify heating, industrial processes, and transportation.

The Energy Information Administration (EIA) data from 2012 to 2022 showed that U.S. energy production increased by 6.4%, and carbon emissions decreased by 23.5%. Over that same period, solar and wind electrical energy production increased from 145 TWh to 639 TWh, and coal decreased from 1,541 to 831 TWh.[52] Reducing coal as a fuel source to generate electricity and increasing the use of wind and solar were significant steps in the clean energy transition.

---

[52] EIA, "Table 1.2: Summary Statistics for the U.S. 2012-2022," *U.S. Energy Information Administration*, (October 19, 2023), accessed June 27, 2024.

The data in Table 3 provides a view of the impact of the clean energy transition on the U.S. electric grid over ten years from 2012 to 2022 by energy source. Wind and solar clean electricity production increased from 3.6% of the total in 2012 to 15.1% in 2022. Coal and natural gas decreased from 69% to 59%. Much effort is needed to achieve a carbon-neutral electric grid.

Table 3: Energy Information Admin. Total U.S. Electrical Energy Production[53]

Source	2012 (TWh)	2022 (TWh)	Change (TWh)	% Change
Coal	1,514	831	-683	
Natural Gas	1,226	1,687	+461	
Nuclear	769	771	+2	
Hydroelectric	276	255	-21	
Wind	141	434	+293	
Solar PV	4	205	+201	
Total All Sources	3,976	4,231	254	+6.4
CO_2 Emissions (M-Tons)	2,157	1,650	-507	-23.5

---

[53] EIA, "Table 1.2: Summary Statistics for the U.S. 2012-2022," *U.S. Energy Information Administration*, (October 19, 2023), accessed June 27, 2024.

Clean Energy Production by Nation

Figure 12a shows Electricity Generation from clean energy technologies (wind, solar, hydroelectric, geothermal, tidal waves, biomass, and nuclear) by country from 1975 to 2022.[54]

Figure 12a: Electricity Generation from Clean Energy Sources[55]

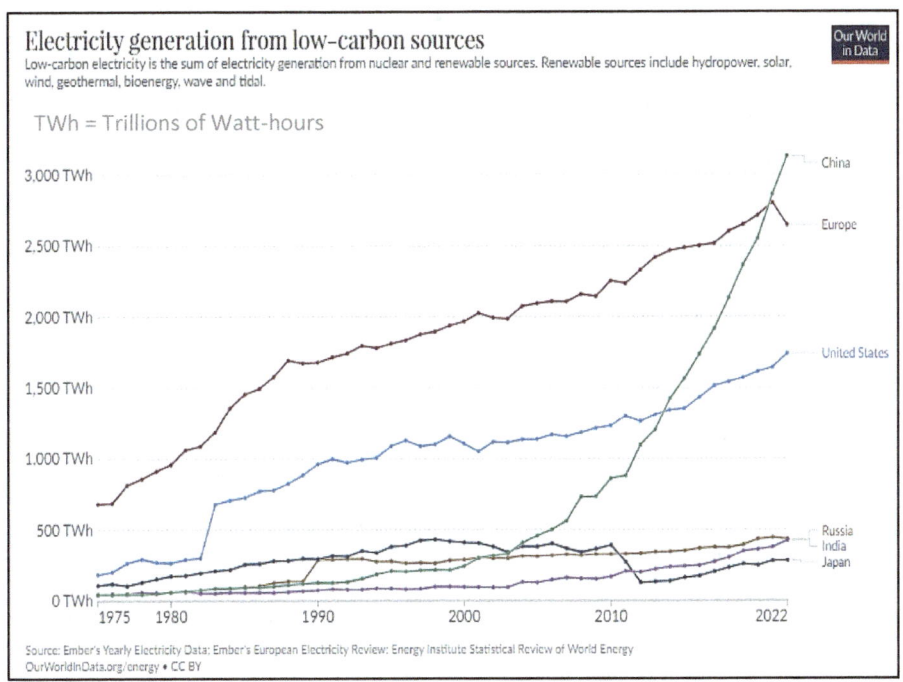

Figure 12b shows the electrical power produced over time by country. In 2018, China became the world's leading producer of wind-sourced electrical power. Figure 12c shows the electrical energy produced over time by country. In 2017, China became the world's leading producer of solar-sourced electrical energy. Europe is second in wind capacity and solar energy production. The U.S. is third in both categories.

---

[54] Ember's Yearly Electricity Data, "Electricity Generation from Low-Carbon Sources," *Our World Data*, (April 2022), accessed September 15, 2022.

[55] Ember's Yearly Electricity Data, "Electricity Generation from Low-Carbon Sources," *Our World Data*, (April 2022), accessed September 15, 2022.

38

Figure 12b: Electrical Power Generation from Wind[56]

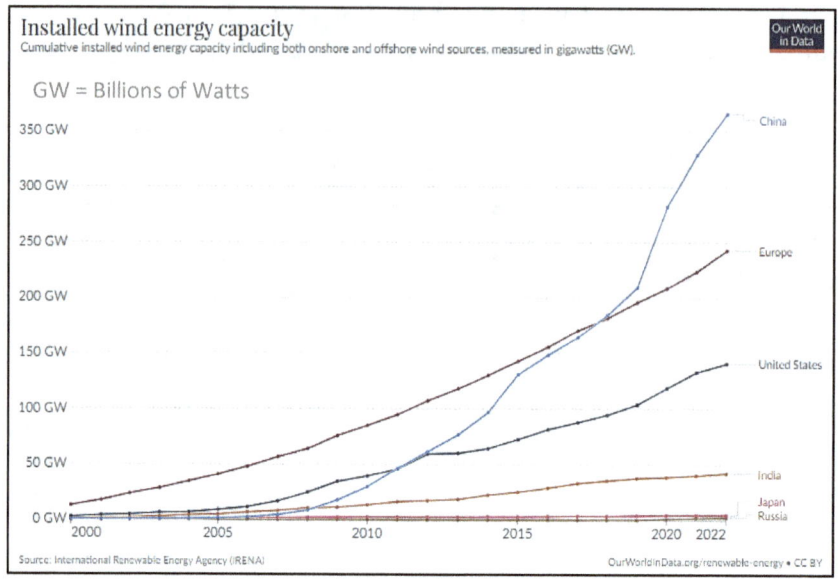

Figure 12c: Electrical Energy Generation from Solar[57]

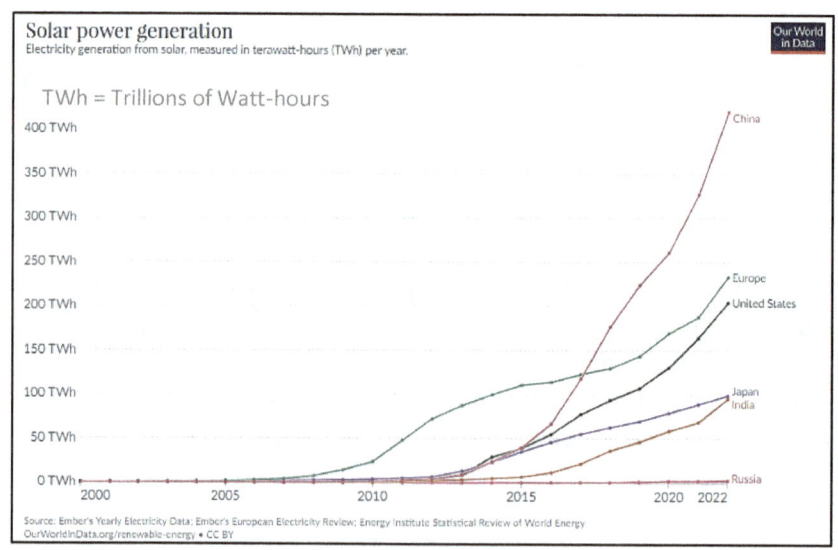

---

[56] Ember's Yearly Electricity Data, "Electricity Generation from Low-Carbon Sources," *Our World Data*, (April 2022), accessed September 15, 2022.

[57] Ember's Yearly Electricity Data, "Electricity Generation from Low-Carbon Sources," *Our World Data*, (April 2022), accessed September 15, 2022.

As shown in Figure 12d, the U.S. Energy Information Administration (EIA) forecasts renewable wind and solar energy to lead the growth of U.S. electricity generation in the next two years. The Short-Term Energy Outlook (STEO) indicates that solar power is expected to grow by 75% and wind energy by 11%, each achieving about 160 to 170 GW of capacity. By contrast, the use of coal is projected to decline by 18%. Nuclear and renewable hydroelectric, biomass, and geothermal will remain constant, while natural gas, at about 495 GW of capacity, will continue to be the largest source of U.S. electrical power generation.[58] Renewable energy expansion has benefited from highly favorable economics. Replacing fossil fuels with low-cost solar and wind-generated electricity lowers utility operating costs and consumer costs.[59]

It is encouraging to note that renewable energy sources have significantly increased, from 200 to 450 GW, indicating a promising future for clean energy. While encouraging progress has been made, fossil fuels remain the leading energy source in the U.S. These data show that the electrical grid's clean energy transition has a long way to go, and the current pace of change is too slow.

Figure 12d: U.S. Annual Electric Generating Capacity (2018-2025)[60]

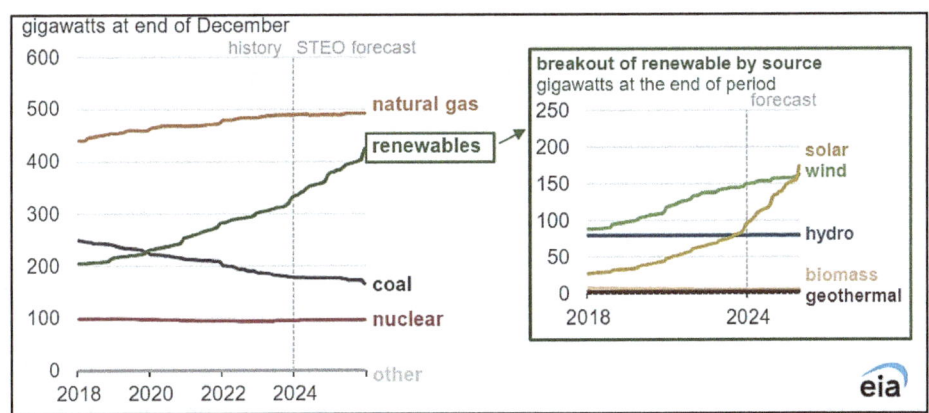

---

[58] EIA, "Solar and Wind to Lead Growth of U.S. Power Generation for the Next Two Years," *U.S. Energy Information Administration*, (January 16, 2024), accessed June 27, 2024.

[59] Marcacci, Silvio, "Renewable Energy Prices Hit Record Lows: How Can Utilities Benefit from Unstoppable Solar and Wind?" *Forbes: Business-Energy*, (April 14, 2022), accessed June 27, 2024.

[60] EIA, "Solar and Wind to Lead Growth of U.S. Power Generation for the Next Two Years," *U.S. Energy Information Administration*, (January 16, 2024), accessed June 27, 2024.

Power Plant Costs Comparisons (LCOE)

In the U.S., the electricity/heat sector is undergoing a "fundamental transformation."[61] Solar and wind are replacing coal as a fuel source. The cost of renewable energy is falling, the amount of wind and solar power is increasing, and carbon emissions from the electricity/heat sector are decreasing.

Figure 13 shows the Levelized Cost of Electricity (LCOE) by energy source. The lower the LCOE, the better the economics of building and operating a power plant.

Figure 13: Levelized Cost of Electricity (LCOE) Generation Nationally[62]

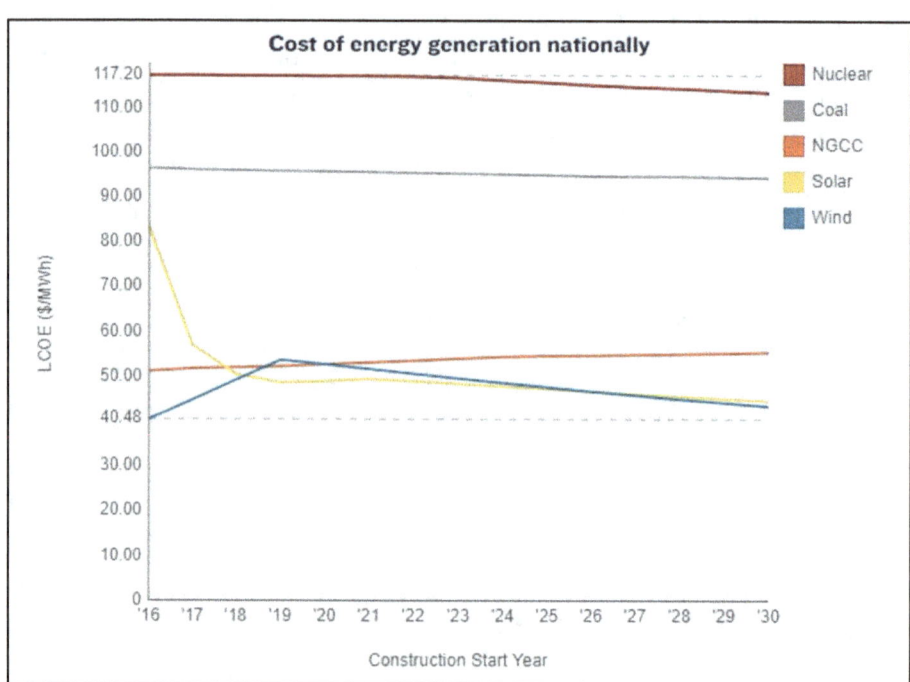

Note: The NGCC natural gas combined cycle uses heat generated from burning natural gas to turn a gas turbine, and the exhaust heat makes steam and runs a steam turbine.

---

[61] The National Resource Defense Council, "Cost of Building Power Plants in Your State," accessed September 15, 2022.

[62] The National Resource Defense Council.

These data show that in 2022, the LCOE for nuclear power will be $117/MWh, coal $95/MWh, natural gas combined cycle (NGCC) $53/MWh, wind $51/MWh, and solar $49/MWh. By bringing solar and wind power plants into production, the electricity/heat sector will benefit from a lower cost of power generation and reduced carbon emissions.

Bringing cost-effective, renewable, sustainable, clean energy, such as wind and solar, into service benefits the transportation sector. Conversion of cars and trucks from internal combustion engines to electric motors would replace gasoline and diesel with electric batteries. Battery systems allow cars and trucks to run without using GHG-producing fossil fuels.

Energy Returned on Energy Invested (EROI)

"Energy return on investment (EROI) is the principal metric of net energy analysis." It "compares the energy output from a technology with the energy that society must invest in delivering that energy."[63]

Energy return on investment is the ratio of useable energy produced by a power plant (e.g., nuclear, coal, natural gas, wind, solar, etc.) divided by the energy invested over the power plant's expected lifetime (this includes the energy needed to build and run the power plant, supply the plant with an energy source, and convert the source energy into useable energy). The EROI is important because energy must be expended to produce energy. The greater the EROI ratio, the greater the amount of usable energy supplied for a given amount of energy consumed, and the greater the benefit to the environment.

EROI is a complicated calculation, and many different approaches have been used. EROI accounts for the efficiency of conversion of the source energy into usable power. The greater the efficiency of conversion, the greater the EROI. For example, as solar panels improve their efficiency in converting sunlight into electricity, the EROI increases. The greater the energy cost of source energy extraction, the lower the EROI. As coal becomes more challenging to find and the cost to extract it increases, its EROI decreases. Including energy for waste disposal reduces the EROI for coal and nuclear significantly. The energy needed for carbon capture and climate change impacts reduces the EROI for fossil fuels.

Table 4 summarizes the EROI estimates of the electricity/heat sector from various energy technologies. The reported EROI values use modern energy analyses. The EROI for fossil fuels has decreased over time due to the increased energy required to find and extract this energy source (e.g., deeper wells). In addition, historical fossil fuel EROI calculations have not included the energy cost of carbon capture, which is needed to achieve the Paris Agreement limit of 1.5°C above preindustrial levels.

The estimated minimum EROI required to run an economy is 7. EROIs below this range result in unacceptably high levels of energy resources and economic activity diverted to producing the energy needed to run the economy. Thus, the EROI value for a given energy technology determines the relative benefit to the economy. The data provided in Table 4 summarizes published

---

[63] Carbajales-Dale, Michael, "Energy Return on Investment," *Science Direct*, accessed August 8, 2023.

EROI values.[64][65][66] The first column does not include carbon capture or GHG costs. The second column adjusts the EROI for carbon capture and climate impacts.

Table 4: Energy Return on Investment (EROI) Comparisons [66, 67, 68]

Energy Technology	EROI without Carbon Capture	EROI with Carbon Capture
Hydroelectric	84	84
Coal	46	18
Natural Gas	20	7
Wind	19	20
Solar PV	16	16
Nuclear	15	15

From the beginning of industrial expansion in the 1850s to today, coal has been the fossil fuel of choice, partly due to its relatively high EROI. The full EROI, which includes the energy cost required to mitigate the climate impact, dramatically reduces coal's EROI. When climate impacts are included in EROI calculation, wind and solar energy technologies become viable alternatives.

The periodic nature of wind and solar technologies requires energy storage technologies during off-peak production and supplemental energy sources without intermittent limitations. Solar and wind costs (LCOE) are lower than those of the nearest fossil fuel neighbor, the natural gas combined cycle (NGCC). The EROI of solar PV will improve with time as conversion efficiencies improve, while fossil fuel EROIs will continue to decline as easily extractable supplies of fossil fuels diminish.

How Do Fossil Fuels, Trees, Solar PVs, Windmills, and Rechargeable Batteries Work?

Fossil Fuels: When fossil fuels are burned, they produce heat through a chemical reaction called oxidation. Methane, CH_4, a natural gas fossil fuel, is

[64] Hall, Charles A., et al., "EROI for Different Fuels and the implications for society," *Energy Policy*, 64 (2014) 141-152), accessed August 8, 2023.

[65] Vikram Solar, "EROI of Solar Energy," (Updated April 21, 2016), accessed August 8, 2023.

[66] Inman, Mason, "How to Measure the true Cost of Fossil Fuels," *Scientific American: Environment*, (updated April 1, 2013), accessed August 8, 2023.

called a hydrocarbon because it has hydrogen (H) and carbon (C). When methane is burned, it reacts with oxygen (O_2) to produce carbon dioxide (CO_2), water (H_2O), and heat energy. This reaction is illustrated in Figure 14a below:

Figure 14a: Oxidation of Hydrocarbons

$$CH_4 + 2O_2 \rightarrow 2H_2O + CO_2 + Heat$$

Fossil fuel hydrocarbons consist of multiple - (CH_2) - groups, which, when burned, are converted to CO_2, H_2O, and heat energy. This reaction tells us that each pound of gasoline we burn produces about 3 pounds of CO_2. A gallon of gasoline, at room temperature, weighs about 6 pounds and produces about 18 pounds of CO_2 when burned.

This calculation does not include the fossil fuels burned to extract the oil, transport it to the refinery, refine it to produce gasoline, transport it to the gas station, and pump it into the gas tank. Including these factors increases the amount of CO_2 produced to 25 pounds per gallon of gasoline burned in the car's engine.[67] Burning gasoline produces a large amount of CO_2. When you purchase a gallon of gasoline, you take ownership of the 7 pounds of CO_2 generated to deliver the gasoline to your gas tank. When you burn that gallon of gasoline, you add another 18 pounds of CO2 to the atmosphere. This makes the total CO_2 emitted per gallon of gasoline to run an automobile engine 25 pounds.

Trees: Trees are important because they use sunlight and chlorophyll to convert carbon dioxide (CO_2) and water (H_2O) into glucose ($C_6H_{12}O_6$) and oxygen (O_2), as illustrated in Figure 14b below:

Figure 14b: Photosynthesis

$$6CO_2 + 6H_2O \ \frac{Sunlight}{Chlorophyl} \rightarrow C_6H_{12}O_6 + 6O_2$$

Trees have enzymes that convert glucose into cellulose, a wood material. Trees are the earth's respiratory system. They take in carbon dioxide and

---

[67] Hawken, Paul, *Drawdown: The Most Comprehensive Plan Ever Proposed to Reverse Global Warming*, editor Hawken, Paul, (Penguin: New York, 2017), 142.

release oxygen. The great thing about photosynthesis is that trees take six carbon dioxide molecules to make one glucose molecule. That ratio is an excellent benefit for the reduction of atmospheric CO_2 levels.

Solar Photovoltaics: Figure 15a shows the basic features of a solar cell. It comprises a few essential parts: a glass lens, a top conductor, a silicon layer, a junction, another silicon layer, and a bottom conductor. These parts combine to convert light into electricity, a process known as photovoltaic conversion. This simple yet powerful technology makes solar panels effective at harnessing the sun's energy.

Figure 15a: Drawing of a Solar Panel Photo Voltaic Cell[68]

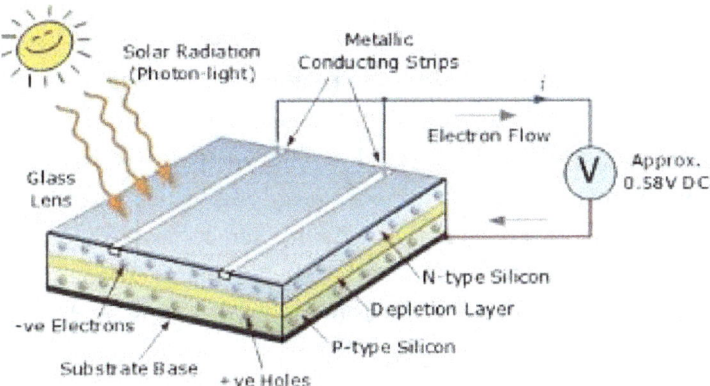

No current flows when the electrons in the silicon semiconductor are in their neutral ground state. When the light energy is absorbed in the junction (depletion layer), it creates energized negative charges (electrons) in the n-silicon and positive charges (electron holes) in the p-silicon. The energized electrons move to the conductor at the surface of the n-silicon. The electron holes move to the conductor at the bottom of the p-silicon. The energized electrons pass through an electrical circuit (to convert direct current into alternating current) and then give up their energy to an electrical device such as a light bulb, toaster, fan, etc. The electrons reach the bottom conductor, combine with the electron holes, and return to their neutral ground state. Thus, the light energy is converted into electrical power by creating energized electrical charges across a semiconductor. The energized electrons travel

---

[68] "Solar Cell Construction and Working Principle," *Electrical Engineering 123,* Accessed October 20, 2023.

through a circuit and then release their power to an electrical device, which may produce light, heat, or mechanical motion.

One of the reasons that solar panels last so long is that they have no moving parts, do not need to undergo chemical reactions, and require no heat input to generate power. The benefit of roof-mounted solar panels is that the electrical power is generated on-site and goes directly into the home or building where the power is needed. This compares to a coal-fired power plant, often located in a remote place where coal must be mined, transported to the plant site, and then burned. The heat from the burning coal is used to make steam. Steam is used to run a turbine. The turbine turns an electrical generator. The electrical power must be transmitted through a grid to a substation and directed to the home or building where needed. Less than 33% of the energy from coal burning goes to running an electrical device. By contrast, over 90% of the electricity generated by rooftop solar panels goes to running an electrical device.[69]

Windmills: Figure 15b shows the essential parts of a three-blade, horizontal-axis windmill. A utility-grade windmill typically has a tower that stands 300 to 800 feet tall with 150 to 350-foot-long blades and can generate 5 to 15 megawatts of power capacity. A computer controls the direction the windmill faces (yaw) and the angle of the blades (pitch). Once the computer program points the windmill in the direction of the wind, the pitch of the blades is adjusted to control the rotation speed. The blades turn at 30 to 60 revolutions per minute (rpm). Gears connect the low-speed turbine shaft to the high-speed generator shaft. The generator, which produces the electricity, turns at about 1,200 to 1,500 rpm. The typical lifespan of a windmill is 20 years.[70] Here, we see how a windmill converts wind energy into mechanical energy (blade rotation) and mechanical energy into electricity (by turning a generator) without burning fossil fuels.

---

[69] "Solar Inverters," *Energy Saving Trust*, (updated 2024), accessed 2/10/2024.
[70] EPA, "Renewable Energy Fact Sheet: Wind Turbines," United States Environmental Protection Agency (updated August 2013), accessed August 10, 2023.

Figure 15b: Schematic of a Three-Blade Horizontal-Axis Wind Turbine (HAWT)[71]

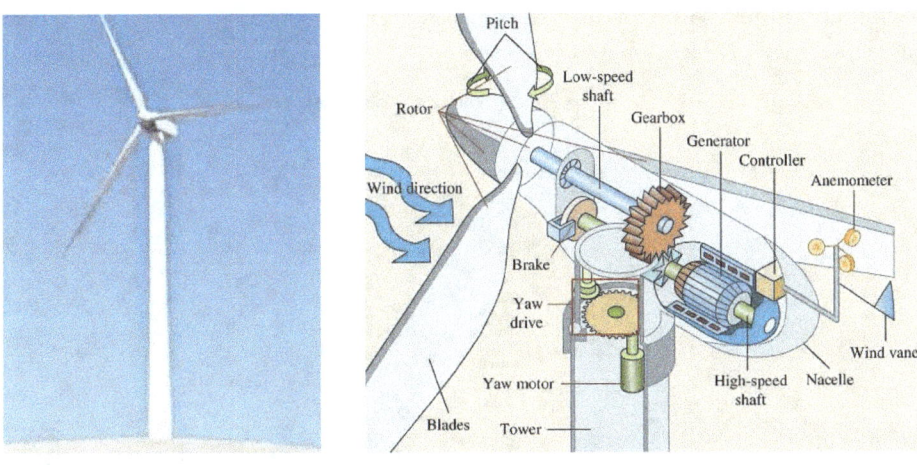

Concerns about wind energy have been addressed. Blades have new designs to reduce wildlife contact and mortality rates. Blade heating systems prevent damage due to ice build-up in cold weather. Wind turbines in less populated areas reduce noise levels and visual aesthetic concerns. Ninety-six percent of windmill components can be repaired, reused, or recycled.[58] Wind has the advantage of requiring essentially no water for power generation.

Wind technology has an excellent LCOE and EROI, and the U.S. has significant wind resources, making wind energy technology a viable option for the U.S. as a renewable, clean, and sustainable energy source. Commercially available wind systems are successfully operating today.

Rechargeable Batteries: Figure 15c shows how a Lithium-ion battery works. Lithium-ion batteries produce electricity through reversible chemical reactions between lithium carbon graphite (LiC_6) and metal oxides (MO_2). These reactions move Lithium ions through a liquid electrolyte, back and forth across the battery's porous separator. When discharging, the LiC_6 anode produces a Li-ion that moves through the separator to react with the MO_2 cathode. When charging, the $LiMO_2$ produces a Li-ion that moves in the opposite direction through the separator to reform LiC_6. To date, the metal-oxide combination that gives the best overall performance (greatest driving

---

[71] "Horizontal-Axis Wind Turbine (HAWT) Working Principle," *Electrical Academia*, accessed August 10, 2023.

48

range, shortest charging time, and longest battery life) in a Li-ion battery is Nickel combined with Cobalt. See the <u>Note</u> below.

Figure 15c: How a Lithium-Ion Battery Works[72]

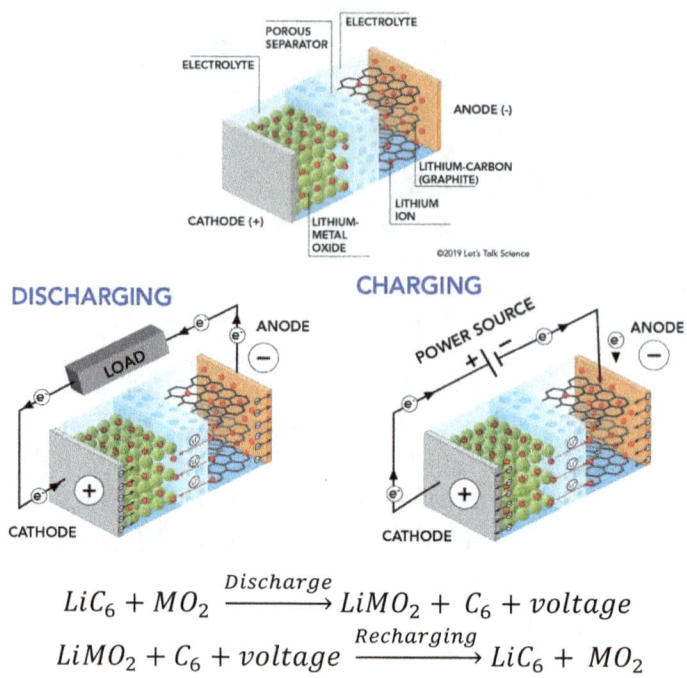

$$LiC_6 + MO_2 \xrightarrow{Discharge} LiMO_2 + C_6 + voltage$$

$$LiMO_2 + C_6 + voltage \xrightarrow{Recharging} LiC_6 + MO_2$$

Automobile Li-ion batteries discharge as they run the electric motor of the EV and charge when connected to a power source. When the battery is charged using a solar PV system on the roof of a home, sunlight provides the power to charge the EV battery. In this case, the electric motor of the EV is running on energy provided by the sun, which produces no CO_2 during charging and running.

Home battery systems connected to solar panels recharge during the day. Excess electricity from the solar PV system would be stored in the electrochemical battery system. During an outage, these home battery systems power critical circuits (e.g., refrigerator, HVAC, Wi-Fi, etc.). This approach is environmentally superior to fossil-fuel-powered generators.

[72] Chapman, Becky, "How Does a Lithium-Ion Battery Work?" *Let's Talk Science*, (September 23, 2019), accessed 11/08/2023.

Note: Later in this book, we will discuss the environmental and social justice issues associated with the mining of Cobalt. Long-term Nickel-Cobalt cathodes are not sustainable because Cobalt is a relatively scarce, high-cost metal. Recent research has demonstrated that low-cost, highly abundant, commercially available Iron (Fe) powder can prepare Iron Phosphate and Iron Fluoride salts to replace the expensive Nickel-Cobalt Oxides.[73] These findings indicate that an Iron cathode battery has a more extended driving range, reasonable charging rates, and acceptable battery life at a lower cost. Finally, researchers have made significant advances in solid-state EV battery technology.[74] Solid-state EV batteries use a solid-state electrolyte, making them lighter and much safer than Li-ion batteries with liquid electrolytes. They will provide a 500-mile driving range and a 10-minute charging time. Automakers are working to reduce manufacturing costs and increase battery life.

Wind as a Clean-Renewable Energy Source in the U.S.

In 2022, 42 states with utility-scale wind power plants generated 434 terawatt hours of electricity, about 10% of the U.S. utility-scale power generation. Figure 16 shows the utility-scale wind electricity generation by each state in 2022. The Central Plain states, "Texas, Oklahoma, Kansas, Iowa, and Illinois… combined to produce 57% of the total U.S. wind electricity generation in 2021."[75]

---

[73] Yu, Mingliang, et al., "Unlocking Iron Metal as a Cathode for Sustainable Li-ion Batteries by an Anion Solid Solution," *Science Advances*, Vol. 10, Issue 21, (May 23, 2024), accessed 7/20/2024.

[74] Osmanbasic, Edis, "Solid-State EV Batteries are closer than you think," engineering.com, (February 22, 2024) accessed 8/2/2024..

[75] The U.S. Energy Information Administration, "Wind explained: Where wind power is harnessed," *Environmental Impact Assessment (EIA)*, (updated April 20, 2023), accessed August 15, 2023.

Figure 16: Utility-Scale Wind Electricity Generation by State, 2021[76]

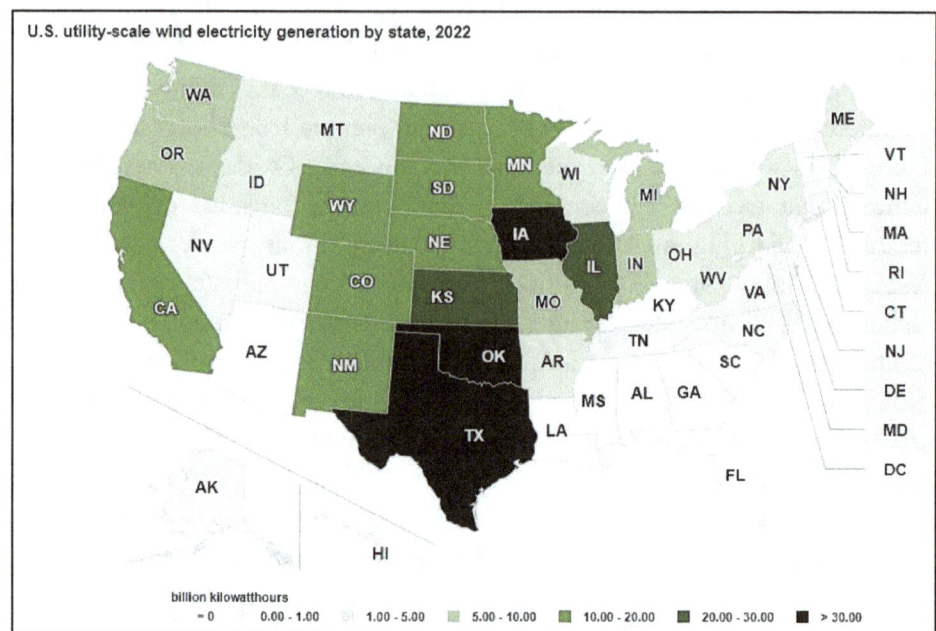

Wind-sourced electrical power generation has great potential to reduce CO_2 emissions. Built with careful planning, utility-scale wind turbines deliver renewable, clean, sustainable, and relatively low LCOE power to the electrical grid. Site location is critical to a utility-scale wind-generated power plant's success. Site evaluation factors include the impact on wildlife (migratory birds), aviation (air traffic control and military missions), urban areas, and national parks. Wind-sourced power plant production capacity and EROI economic feasibility depend on wind quality. Good places for utility-scale (>1 megawatt) wind power plants are in locations where the average wind speed is 13 miles per hour (>5.8 meters per second). Favorable sites in the U.S. include open plains, coastal regions, and mountain gaps.

Figure 17a illustrates the acceptable average wind speeds in the U.S. at a 100-meter (330-foot) hub height. The highest quality wind resource for utility-grade power generation is located at higher elevations in mountain gaps, where the wind intensity and consistency are enhanced due to a 'funneling' effect. These regions have wind speeds suitable for power generation around the clock.

---

[76] The U.S. Energy Information Administration.

The electricity produced during the day, when there is peak power demand, can be seamlessly integrated into the grid. At night, when electricity demand decreases, the excess wind-sourced electricity can be used to pump water for hydroelectric power generation or diverted to electrolyzers for hydrogen production. In the future, hydrogen could serve as a carbon-free fuel source for transportation, further enhancing the versatility of wind energy.

Figure 17a: U.S. Annual Average Wind Speed at 100 Meters[77]

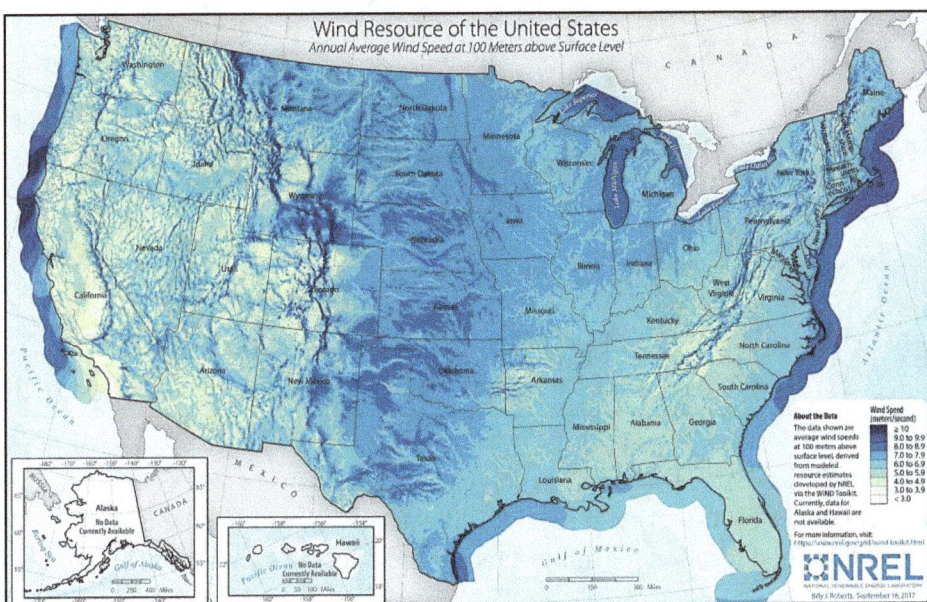

Figure 17b indicates the average wind speed at an 80-meter (260-foot) hub height; these data indicate acceptable wind quality for power generation throughout the central plain states.

---

[77] The Office of Energy Efficiency & Renewable Energy.

Figure 17b: U.S. Average Annual Wind Speed at 80 meters[78]

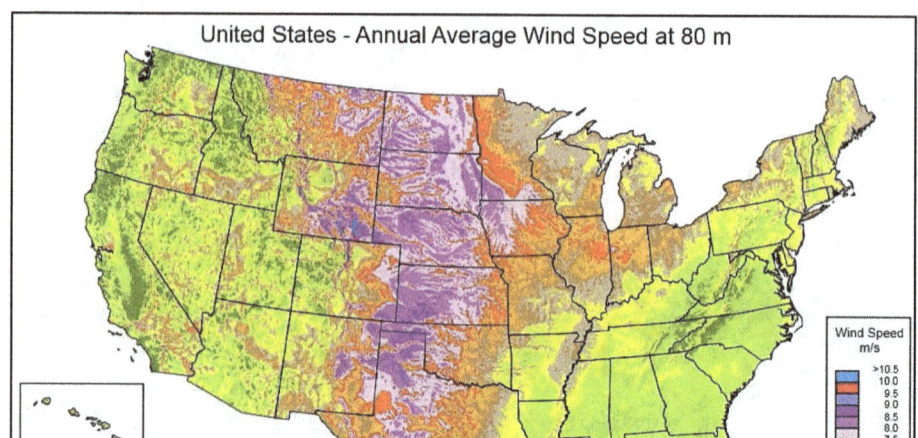

Figure 17c indicates acceptable average wind speeds in the U.S. coastal regions at a 90-meter (300-foot) hub height. The shallow waters off the eastern coast provide a good location for utility-grade wind-sourced power plants.

---

[78] The Office of Energy Efficiency & Renewable Energy, "Wind Energy Maps and Data," *Wind Energy Technologies Office – Wind Exchange*, accessed September 15, 2022.

Figure 17c: U.S. Annual Average Offshore Wind Speed at 90 meters[79]

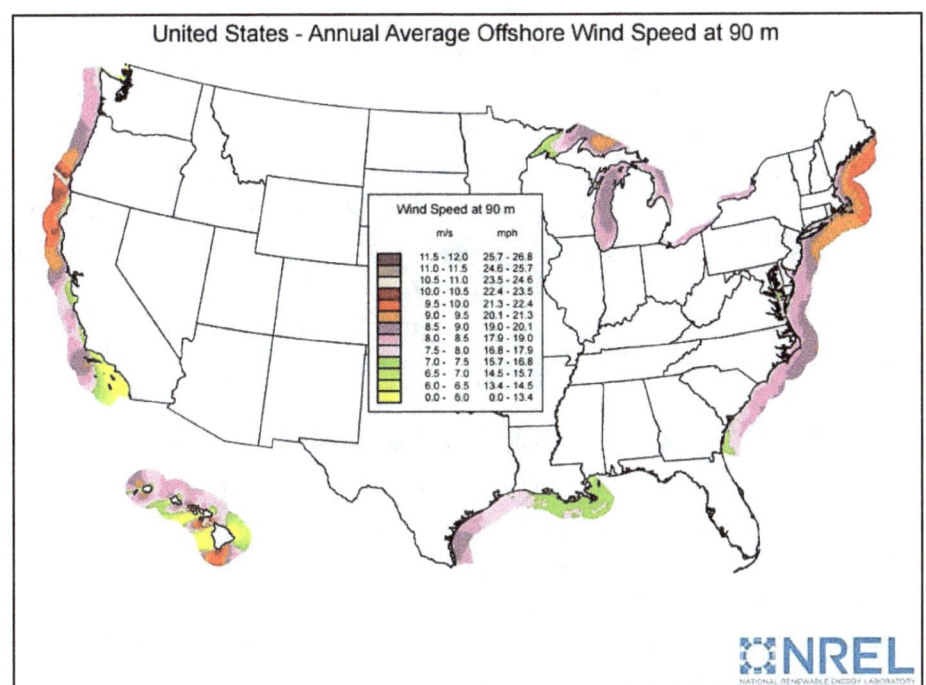

The U.S. has two offshore wind energy utilities in operation: a 30 megawatt (MW) wind farm off the coast of Rhode Island and a 12 MW capacity project off the coast of Virginia.[80] The Coastal Virginia Offshore Wind (CVOW) project plans to install a 2,600-megawatt wind farm by 2026, making it the largest offshore wind farm in the U.S.[81] The wind farm will consist of 176 14.7 MW windmills, each with an 800-foot hub height and 354-foot blades, delivering 9.5 million MWh per year of clean energy. The windmills must only rotate 40% of the time to meet this power generation goal. The windmills will be located out of sight, 27 miles off the coast of Virginia, in 80-foot water. These windmills are designed to withstand sustained hurricane-force winds of 112 miles per hour and gusts up to 157 miles per hour. The windmill design

[79] The Office of Energy Efficiency & Renewable Energy.
[80] The U.S. Energy Information Administration, "Wind explained: Where wind power is harnessed," *Environmental Impact Assessment (EIA)*, (updated March 30, 2022), accessed September 15, 2022.
[81] Dominion Energy, CVOW Project, (copyright 2023), accessed October 28, 2023.

will minimize the impact on migratory birds and marine mammals, and the windmill parts are recyclable. The windmill location will not interfere with commercial shipping or military operations. This project will generate $3B in fuel savings for customers. It will provide 1,100 good-paying, clean-energy jobs, economic development opportunities, and tax revenues. Dominion Energy, Virginia's largest utility, seeks to reach net-zero emissions by 2050.[82]

Wind energy, which has a low cost (LCOE) and high return (EROI), provides the U.S. with tremendous potential for electricity generation. The electricity and heat sectors generate roughly 32% of the U.S. GHG emissions. Transitioning from fossil-fuel-powered electricity and heat generation to clean energy would significantly reduce GHG emissions.

Solar PV as a Clean-Renewable Energy Source in the U.S.

U.S. solar PV power comes from utility-level solar farms, and local generation primarily focuses on rooftops and community solar projects. Solar power comes from thermal generation, which uses mirrors to concentrate sunlight, produce heat, and generate electricity.

In 2022, solar PV power plants generated 202 terawatt hours (TWh) of electricity, about 5% of the U.S. utility-scale power generation. About 141 TWh from utility-grade solar PV and 61 TWh from local and community scale.

In the renewable energy sector, solar PV has the fastest projected growth rate in the U.S. The expansion of solar will accelerate the U.S. clean energy transition, allowing a greater rate of reduction of fossil fuel consumption for electricity.

Figure 18 shows the top 10 states' solar PV capacity in 2022.[83] These states combined to produce about 50% of the electricity generated in the U.S. by solar PV in 2022.

---

[82] Comay Laura B., and Clark, Corrie E., "Offshore Wind Energy: Federal Leasing, Permitting, Deployment, and Revenues," Congressional Research Service, (updated December 7, 2021), accessed September 15, 2022.

[83] Fernandez, Lucia, "Leading States in Solar Photovoltaic Capacity in 2022," *Statista: Environment and Energy*, (October 20, 2023) accessed June 27, 2024.

Figure 18: Top 10 States' Solar PV Capacity (MW) in 2022[84]

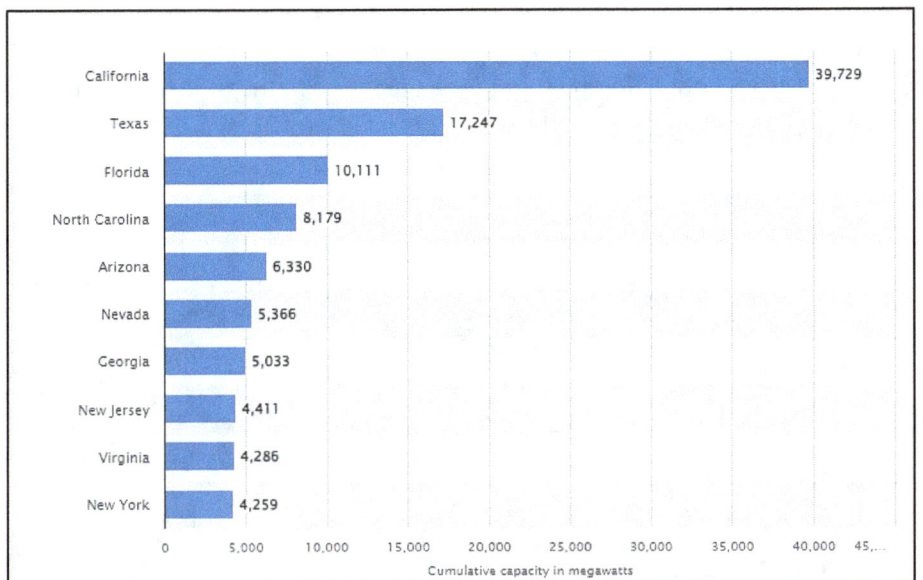

In 2018, California set a goal of 100% clean electric power by 2045. In 2022, the Inflation Reduction Act extended tax credits for solar projects and provided funds to states, community groups, residential, non-residential, schools, and non-profit organizations for solar installations. In 2009, the American Recovery and Reinvestment Act funded U.S.-based photovoltaic research and solar panel manufacturing to increase green jobs and the U.S. competitive advantage in solar PVs.

---

[84] Fernandez, Lucia, "Leading States in Solar Photovoltaic Capacity in 2022," *Statista: Environment and Energy*, (October 20, 2023) accessed June 27, 2024.

Figure 19 illustrates the annual average daily solar resource potential for the U.S.[85] This map was generated by measuring total irradiance from the sun on a horizontal surface; the solar resource data was collected from 1998 to 2016. The map accounts for the average changes due to the time of year and the effects of cloudy days and weather events, and shows that the Southwest and Hawaii have the most significant solar potential.

Figure 19: The U.S. Annual Solar Resource Potential[86]

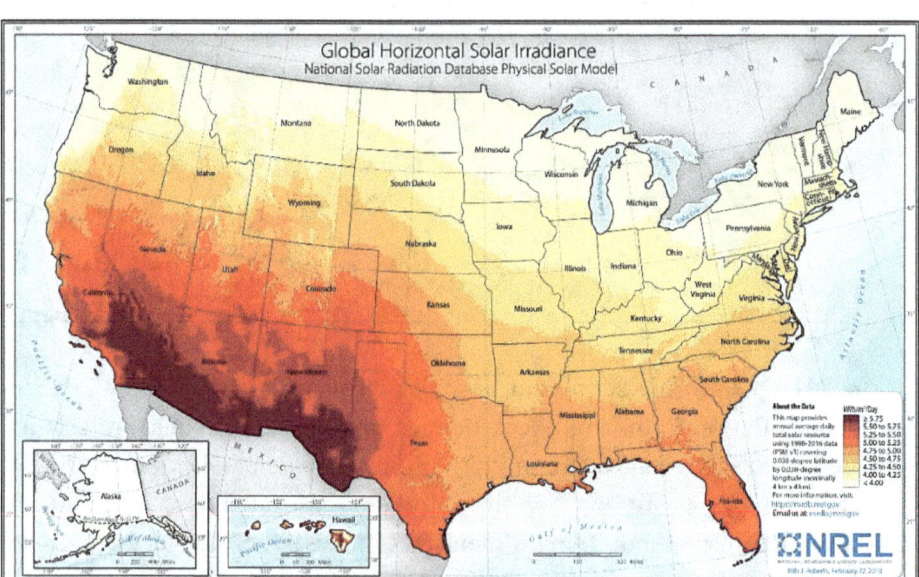

Solar panels have a 30-year lifetime. They contain glass, silicon, aluminum, and small amounts of indium, selenium, and cadmium, which must be appropriately disposed of. Researchers have begun developing end-of-life processing of solar panels. The EPA has proposed new rules for processing end-of-life solar panels and lithium batteries. The EPA program aims to responsibly recycle and "return valuable and critical minerals to the economy."[87]

---

[85] Sengupta, M., et al., "The National Solar Radiation Data Base (NSRDB)." Renewable and Sustainable Energy Reviews 89 (June 2018): 51-60.

[86] Sengupta, M., et al., "The National Solar Radiation Data Base (NSRDB)." Renewable and Sustainable Energy Reviews 89 (June 2018): 51-60.

[87] EPA, "Improving Recycling and Management of Renewable Energy Wastes: Universal Waste Regulations for Solar Panels and Lithium Batteries," *United States*

Electric Vehicles for Transportation[88, 89]

The transportation sector generates roughly 30% of the U.S. GHG emissions. The transition of automobiles and light trucks from fossil-fuel-powered vehicles to EVs and clean energy to charge batteries would significantly reduce GHG emissions. Electric vehicles (EVs) have a rechargeable (Lithium-ion) battery that supplies energy to an electric motor and produces no tailpipe emissions. The transition of the transportation sector from gasoline-powered engines to electrically powered motors is a critical component in reducing carbon emissions in the transportation sector. Converting automobiles and light trucks to EVs is necessary to meet COP26's goal of net-zero GHG emissions by 2050.

Several myths about EVs have been promoted to question the climate benefits of electric vehicle technology. A recent EPA study addressed these myths.[90]

1. Myth: EVs emit more GHG than ICEs because of power plant emissions. Fact: EVs have much lower GHG emissions than internal combustion engines (ICEs)—even when accounting for emissions from electricity-generating power plants. The GHG benefit of EVs becomes greater when charged by electricity from a clean energy source like wind or solar.

2. Myth: Battery manufacturing makes EVs worse for the environment than ICEs. Fact: Over the lifetime of the vehicle, EVs emit ~60% fewer GHG emissions than ICEs, even when including battery manufacturing and charging, which makes them better for the climate than vehicles that burn fossil fuels to power internal combustion engines.

3. Myth: The increase in EVs will collapse the U.S. power grid. Fact: The electric grid has sufficient capacity to charge EVs as they come into use in the near term. In the long term, the electric grid will need to add capacity when most vehicles on the road are EVs. Plans and funding

---

Environmental Protection Agency, (updated December 14, 2023), accessed June 27, 2024.

[88] EPA, "Electric and Plug-in Hybrid Electric Vehicles," *United States Environmental Protection Agency*, (updated August 28, 2023), accessed August 29, 2023.

[89] Alternative Fuels Data Center, "Electric Vehicle Charging Stations," U.S. Department of Energy, accessed September 17, 2024.

[90] EPA, "Electric Vehicle Myths," *United States Environmental Protection Agency*, (updated August 28, 2023), accessed August 29, 2023.

have begun to upgrade the electric power grid. Wind and solar energy sources will add renewable, clean energy capacity to the electric grid, further reducing EV GHG emissions.

4. Myth: There is nowhere to charge EVs. Fact: Currently, the U.S. has 51,000 charging stations available to the public. Plans and funding have begun to build a network of EV charging stations along interstate highways, communities, and neighborhoods.

5. Myth: EVs do not have the range to handle trip demands. Fact: Current EV batteries typically run 200 miles or more between charges. This range is more than adequate for those who run primarily short trips. Batteries can be recharged at home overnight. EV manufacturers are developing new models with significantly longer driving ranges. The EV battery range is temperature-dependent. Cold weather and using the car heater may reduce the battery range by as much as 40%.

6. Myth: EVs take too long to charge. Fact: Battery charging time is dependent upon the source voltage. A 480-volt charging station can recharge a battery in one hour. Phone apps provide the location and availability of charging stations. EVs come with a GPS that can guide you to charging stations. Federal laws have called for adding new charging stations along all interstates. With these technologies, trips can be planned accordingly.

7. Myth: EVs are not as safe as ICEs. Fact: EVs must meet all safety standards that ICEs must meet. EV battery packs must also meet testing standards and have additional safety features that shut down the electrical system when a collision is detected.

Conclusion

The U.S. clean energy transition from fossil fuels to renewable energy has reduced CO_2 emissions and improved the economy. From 2000 to 2022, per capita emissions in the U.S. have decreased from 20.7 to 14.9 metric tons per person per year.[91] Total U.S. CO_2 emissions have decreased from 6 billion to 5 billion tons.[92] Moreover, the U.S. per capita gross domestic product has grown

---

[91] Ritchie, Hannah, and Roser, Max.
[92] Ritchie, Hannah, and Roser, Max.

yearly from \$50,000 to \$67,000 per person.[93] The energy transition makes it economically possible to reduce CO_2 emissions and grow the economy. This movement is in the right direction, but...

As mentioned at the beginning of this section, the global transition to clean energy has been moving too slowly. The technologies presented here are demonstrated and available. To achieve the Paris climate limits of 1.5 to 2°C temperature rise above pre-industrial levels by 2100, we must accelerate clean energy development and use clean and renewable energy sources to electrify heating, industrial processes, and transportation.

The technologies developed, demonstrated, and successfully implemented must be utilized more widely to continue reducing these emissions in the years to come to avert the worst consequences of human actions contributing to climate change.

---

[93] Real Gross Domestic Product Per Capita, *FRED Economic Data*, (updated Jan. 25, 2024), accessed 2/10/2024.

Chapter 4. Economics, Populism, and the Cultural Divide

High-quality data (current and historical), numerous reproducible and independent scientific analysis methods, and highly accurate computational models have produced a statistically significant, quantifiable correlation between human activity and climate change. Modern scientific research by more than 2,000 scientists in over 20,000 peer-reviewed journals has resulted in 99.9% agreement that climate change is real and human-caused.[94] The impact of these scientific results "has [helped to shift] the climate debate [from] arguments about whether climate change is real and [human]-made to focus on what […we] should do about the threat."[95]

Many in the U.S. claim the economy is of much greater importance than climate change. Here, we will discuss how economics, politics, and the cultural divide have influenced our understanding of and response to climate change. We will show that many contend that taking climate action *and* maintaining a robust economy with strategic energy independence is impossible. Many hold that these climate actions would harm economic prosperity and require burdensome personal sacrifice. We will discuss how many downplay the severity of the damage caused by climate change, ignore peer-reviewed findings from verifiable data, and promote pseudoscience and fake experts to discredit actual experts.

Economics and the Environment

Economist Jonathan Park notes that the sharp increase in fossil fuel consumption, which began in the mid-1800s, corresponded to capitalist economic expansion in Western Europe and North America: "Capitalism was designed as a mechanism for efficiently allocating scarce resources, encouraging human ingenuity, and improving the quality of life for those willing and able to participate in the system."[96] This economic model, coupled with the development of new technologies, enabled the conversion of raw materials into commercially available commodities, the sale of which created

---

[94] Ramanujan, Krishna, "More than 99.9% of studies agree: Human-caused climate change," *Cornell Chronicle*, (updated October 19, 2021) accessed August 12, 2023.

[95] Freeman, Jody, Guzman, Andrew, "Climate Change and U.S. Interests," *Columbia Law Review*, 109, no. 6 (Oct. 2009): 1531.

[96] Park, Jonathan T., "Climate Change and Capitalism," *Consilience: The Journal of Sustainable Development*, 14, no. 2 (2015): 189.

economic growth and prosperity. The tremendous financial success of the capitalist system has benefited society in many ways, but has also produced some unwanted consequences. The extraction and conversion of natural resources into disposable products have generated toxic waste streams that have polluted the environment and harmed ecosystems.

> *"Global climate change has been the most severe consequence of our society's excessive atmospheric pollution… Anthropogenic climate change poses an imminent threat to the planet's life-sustaining ecological systems, and it represents one of humanity's most difficult challenges to date."*[97]

To solve this problem, significant reductions in fossil fuel consumption are needed. These reductions can only occur through a global commitment to sustainable natural resource consumption and conversion to alternative, non-fossil-fuel-based energy sources.

The capitalist economic system promotes cash flow, profit growth, wealth generation, and ownership. The problem is unpriced externalities.[98] Our current capitalist system does not value environmental protection and emission reduction enough. Governments, businesses, and consumers can make valid economic justifications for shifting to clean energy supplies.

Consumer spending comprises roughly 70% of the US gross domestic product. The remainder is business investment, government spending, and exports;[99] consumers have a significant role in directing the sustainable and fair transition to a clean energy economy. Government incentives and targeted regulations kickstart the process, help overcome the natural resistance to change, and guide consumers to shift to clean energy supplies and electric vehicles for personal transportation. These changes will significantly reduce the CO_2 emissions from home electricity/heat and the transportation sectors. Corporations in a free market competitive environment will respond to these shifts in consumer purchases with business investments and innovation that

---

[97] Park, Jonathan T., 189.

[98] Garen, David C., "Mennonite Values in a Warming World," *Anabaptist World*, (May 20, 2019), accessed) March 4, 2024.

[99] Amadeo, Kimberly, "Components of GDP Explained: Four Critical Drivers of America's Economy," *The Balance*, (updated January 18, 2022) accessed September 15, 2022.

produce good-paying jobs and corporate profits. It is possible to imagine a future where the economy remains stable through the transition and has a positive outlook.

Alternatively, some argue that phasing out fossil fuels and converting them to clean energy sources will increase production costs, increase prices, or reduce profits. Some politicians downplay the warnings of a climate catastrophe based on the argument that taking "draconian" environmental action will ruin the economy. One Senator recently told the New York Times, "I do not want to be lectured about what we need to do to destroy our economy in the name of climate change."[100] Voters are concerned about the economy since it gives people their livelihoods and maintains living standards.

An apparent tension exists between corporations' quarterly profits and dividend delivery to stockholders versus long-term climate change concerns. A powerful, experienced U.S. senator recently said he wants to hold hearings on Electric Vehicle (EV) batteries. He wants answers to supply chain questions, alternatives to EVs, an impact study on shifting from internal combustion engines (ICEs) to EVs, and an answer to the question, "What do we do with the oil industry?"[101] One answer to this question is making reusable and recyclable products from oil rather than burning oil as a fuel source. The Senate hearing would question the feasibility and merits of transitioning the electricity, heat, and transportation sectors from fossil fuels to renewable energy sources. The fossil fuel industry has lobbied for decades against climate action by funding politicians, advertising, and using media to argue that action is too expensive. The fossil fuel industry spreads confusion and doubt, suggesting that wind, solar, and battery power are worse for the environment than fossil fuels.[102]

The U.S. delegation to the 2021 COP26 conference in Glasgow, Scotland, "set ambitious targets for carbon-free electricity by 2035, and net-zero emissions by 2050."[103] These high ambitions were not supported by the

---

[100] Dorman, John L., "No Lectures on How to 'Destroy' Economy Over Climate Change," *Business Insider*, Jul 23, 2022, accessed August 1, 2022.

[101] Shepardson, David, "U.S. Senator to hold EV battery hearing if GOP takes control," *Reuters*, (October 19, 2022), accessed October 20, 2022.

[102] Martinez, Chris, et al, "These Fossil Fuel Industry Tactics Are Fueling Democratic Backsliding," *Center for American Progress (CAP)*, (Dec. 5, 2023), accessed June 17, 2024.

[103] Jackson, Margaret, and Strauss, Zachary, "Raising US Climate Ambition in Advance of COP26: An Economic and National Security Imperative," Atlantic Council (2021): 1.

conservatives in the U.S. Congress nor by the fossil fuel industry. Over 25% of the delegates to COP26 were from the energy industry! Their lobbying efforts were successful in changing the language of the accord. For example, the commitment to "phase out" coal usage was changed on the last day to "phase down."[104] COP26 agreements fell well short of the 2050 net-zero-emission goals to mitigate the climate catastrophe.[105] Greta Thunberg called COP26 "a global north greenwash festival… a two-week-long celebration of business as usual."[106]

Near-term business and economic goals often take priority over long-term outcomes. In a well-functioning system, the decision-making process balances short-term goals and long-term consequences. Instead, long-term concerns are recast into a threat for the short term and thus projected as a liability with dire consequences. This narrative must be corrected. It is possible to rebalance the system while maintaining acceptable near-term benefits and reducing the long-term adverse effects. To accomplish this in the capitalist system, corporation leaders, government officials, economists, environmentalists, consumers, and citizen groups need a shared voice at the table. For the sake of all stakeholders, a "win-win" environmentally sustainable and just economic redirection must be developed and implemented.

Economic models developed by Nordhaus and Yang[107] compared "cooperative" and "non-cooperative" strategies for reducing CO_2 emissions. They found that international policies that followed a "cooperation" model more effectively reduced global CO_2 emissions than a "non-cooperative" approach.[108] Modeling the economic controls needed to reduce emission levels required the most economically advanced nations to make the most significant changes; "From everyone who has been given much, much will be demanded"

---

[104] The Associated Press, "At COP26, nations strike a climate deal with coal compromise," *National Public Radio*, (November 13, 2021), accessed December 6, 2021.

[105] Guterres, Antonio, "UN Chief calls for action to put out '5-alarm global fire'", *UN Climate Action* (January 2022), accessed February 10, 2022).

[106] Kottasova, Ivana, and Picheta, Rob, "Greta Thunberg slams COP26 as a 'failure' at youth protest in Glasgow," *CNN* (November 5, 2021), accessed December 6, 2021.

[107] Nordhaus, William D., and Yang, Zili, "Regional Dynamic General-Equilibrium Model of Alternative Climate-Change Strategies, *The American Economic Review*, 86, no. 4 (Sep. 1996):741.

[108] Nordhaus, William D., 741.

(see Luke 12:48). A sign of hope for the U.S. clean energy transition was the passage of the Build Back Better Act in 2021. This was a response to the climate crisis and an attempt to keep pace with China's clean energy technology advancements (EVs and solar PVs in particular) so as not to miss out on the clean energy transition's environmental benefits and economic growth opportunities.

According to Gardiner, the complexity of the climate change problem "poses substantial obstacles to our ability to make the hard choices necessary to address it... Even if the difficult questions could be answered, we might still find it difficult to act."[109] The complexity of climate change "makes us extremely vulnerable to moral corruption... [and] easy to engage in manipulative self-deceptive behavior."[110] Corporations doubt scientific warnings and promote inaction as the most desirable option to protect economic interests. At the same time, governments give the false appearance of taking necessary steps by negotiating "weak and largely substanceless global accords... and then heralding them as great achievements."[111]

Several scenarios have been developed that predict climate change will contribute to war and global trade disputes. These conflicts will result from climate-change-induced sea-level rise, reduced food production, economic disruptions, and population migrations. Holms-Dixon stated, "Waves of environmental refugees [...will have] destabilizing effects on... international stability."[112]

Many in the U.S. who oppose acting on climate change have developed scenarios that forecast the U.S. as a climate change winner. Although these arguments are flawed, they have gained the appeal of far-right groups. These arguments falsely claim that maintaining the current fossil fuel consumption rates and U.S. standard of living comes without any economic downside for the U.S. "This argument fails to account for the full spectrum of external costs that climate change will impose on the United States."[113]

---

[109] Gardiner, Stephen M., "A Perfect Moral Storm: Climate Change, International Ethics and the Problem of Moral Corruption," *Environmental Values*, 15, no. 3 (August 2006): 398.

[110] Gardiner, Stephen M., 398, 408

[111] Gardiner, Stephen M., 408.

[112] Homer-Dixon, Thomas F., "On the Threshold: Environmental Changes as Causes of Acute Conflict," *International Security*, 16, no. 2 (Fall 1991): 77. 76-116.

[113] Freeman, Jody, and Guzman, Andrew, 1534.

Populist Opposition

The distribution of wealth and political power, cultural identity, and immigration have become essential themes that have helped mobilize the growing populist movement over the past decade. Several factors have been identified that contribute to populist resistance to climate action. Government climate policies originating at international meetings, such as COP26, are seen as elitist and a benefit to global financial interests. Populist organizations tend to hold anti-elitist and isolationist positions and oppose policies that they fear might result in social and economic changes that threaten cultural norms and produce financial hardship.

Jonathan White's recent work on the populist issues regarding climate change indicates that:

> *"Populism finds resonance in the critique of political necessity and prospers in situations of emergency where policy is rationalized in these terms. As climate change comes to be framed as an emergency to which governments must respond by imposing policies, it becomes a natural target for populist critique."*[114]

Climate change evaluation requires the expertise of professionals in diverse fields of science, often perceived as elitist by some. Populists, in their anti-elitist stance, challenge the authority of experts and assert traditional wisdom and cultural values over elitist approaches. They believe that governments have framed climate change as a crisis so that they can require urgent action. Populists interpret the government's response to the climate crisis as a means to justify taking away their freedoms with mandates and regulations. The clean energy transition will produce unwanted societal changes while leaving the real threat of uncontrolled migration unaddressed.

The Build Back Better Act partially addressed populist economic concerns with promises of good-paying U.S. jobs, inflation reduction, infrastructure development, and solar and wind energy expansion to reduce dependence on foreign oil. However, the populists primarily see Build Back Better as a

---

[114] White, Jonathan, "What Makes Climate Change a Populist Issue," *Grantham Research Institute on Climate Change and the Environment*, (September 14, 2023), accessed June 20, 2024.

government-mandated program imposing a costly clean-energy transition and failing to address their concerns about immigration.

Populists have resisted elitist agendas, government policies, and programs promoting net-zero emissions. They have recast the climate crisis as a threat to their way of living by conflating immigration and climate change into an econationalist agenda. This approach raises fears that climate change will result in uncontrolled migration. It claims that the best government response to climate change is building stronger borders, not a clean energy transition.

The effects of climate change will have a more severe impact on developing nations. Reduced capacity to mitigate harsh economic conditions and prevent food shortages in developing countries will trigger many desperate people seeking to immigrate to more prosperous, developed countries. "This mass upending of lives is set to cause internal and external conflicts that the Pentagon, among others, has warned will escalate into violence."[115]

Conflating climate change with fears of uncontrolled immigration "is a narrative that has flourished in [populist] movements in Europe and the U.S. and is spilling into the discourse of mainstream politics."[116] A conservative politician from the U.K. explained that the reason for addressing climate change "could be found in the fall of the Roman Empire… He argued that the collapse of the Roman Empire was largely […the] result of uncontrolled immigration—the empire could no longer control its borders, people came in from all over the place."[117]

The populist political strategy is shifting from climate change denial to using it as an argument to support its anti-immigration agenda. In the U.S., some politicians are concerned that younger voters are repulsed by climate change denial. The old strategy of discrediting science is no longer working with younger voters who foresee the impact of future climate catastrophes on their lives. The far right is recasting climate denial into econationalism to motivate its base, turning the environmental crisis into an immigration crisis. The econationalism response to the climate crisis is to defend the nation from the dangers of immigration by building strong borders. From the econationalism point of view, the environmental threat is caused by

---

[115] Milman, Oliver, "Climate denial is waning on the right. What's replacing it might be just as scary," *The Guardian*, (Nov. 21, 2021), accessed 12/6/2021.

[116] Milman, Oliver, ibid.

[117] Milman, Oliver, ibid.

overpopulation and excessive pollution in the poorer, developing nations. This false narrative is used to exonerate wealthier nations of complicity in the global climate crisis.

Although wealthy nations and consumer-driven, fossil fuel-burning economies are directly responsible for the greatest sum-total amount of CO_2 emissions, some are attempting to blame environmental destruction on the internal behavior of the poorest nations. Racist tropes characterize the developing nations as a danger to Western economic prosperity and a threat to national security; strong borders, so the argument goes, are needed to prevent massive eco-migrations.

Conservatives blame corrupt liberal media for spreading hysteria about climate change and promoting outrageous climate change policies. They maintain that the government and media elites are not to be trusted. Far-right econationalists proclaim a populist perspective that claims government elites do not comprehend the impact their programs to reduce global CO_2 emissions will have on ordinary people's lives. Recent conspiracy theories falsely claim that "Covid is a dry run for restrictions that governments want to impose with the climate emergency."[118]

The populist movement has begun to successfully take fringe ideas, make them mainstream, and use them to form a national populist political agenda that rejects science, condemns government and media elites, and portrays liberal ideas as destructive to the nation. The econationalists make the false claims that the solution to the climate crisis is more border walls to prevent massive migration, fewer environmental regulations that cost jobs and reduce U.S. competitiveness, decreased government taxes that take disposable income out of people's pockets, elimination of government restrictions that take away personal freedoms, and less spending on social programs that encourage immigration and increase the national debt.

Climate Change and the Cultural Divide

The U.S. culture wars and political divide have helped shape people's attitudes toward climate change. Factors influencing people's risk assessment due to climate change and the perceived probability of occurrence have been studied. According to Klein, Yale researchers have shown that an individual's cultural point of view, informed by their ideological stance, predicts one's assessment

---

[118] Milman, Oliver, ibid.

68

of the risk and probability of climate change more than any other factor.[119] Klein reports that researchers determined that 69% of people with "strong 'egalitarian' and 'communitarian' worldviews" characterized by concerns for "social justice" and "suspicion about corporate power" accept the scientific findings on climate change.[120] Alternatively, 89% of people with "strong 'hierarchical' and 'individualistic' world views" characterized by "support for the industry" and disdain for "government assistance for the poor" reject the scientific conclusions on climate change.[121]

1. Climate change deniers: Many climate change deniers reject the overwhelming preponderance of scientific data identifying human burning of fossil fuels as the root cause of increased CO_2 levels and climate change. They promote non-human causes instead.

Possible alternate non-human causes for climate change have been examined. Scientists have shown that non-human causes of climate change, including sun cycles, changes in the Earth's orbit, and volcanic eruptions, have a relatively minor impact on the Earth's climate. Other human activities, including ozone and aerosol production, have been shown to have a negligible effect compared to the GHGs' influence on the average global temperature change. Hayhoe states that those opposed to the truth about climate change use what scientists call "zombie arguments because they just won't die, no matter how often or how thoroughly they are debunked."[122]

Scientists have thoroughly analyzed the human and natural factors that impact the Earth's climate. Figure 20 plots the global temperature from 1880 to 2017. These data show that the "Observed" average global temperature has increased by about 1.1°C since 1880, a significant rise. Reidmiller states, "The long-term global warming trend observed over the past century can only be explained by human activities' effect on the climate."[123] The black line in each plot shows the "Observed" surface temperature increase above the average value from 1880 to 1910. This means that in 2017, the average global surface temperature was 1.1°C warmer than the average temperature from 1880 to 1910.

---

[119] Naomi Klein, *This Changes Everything*, (New York: Simon & Schuster, 2014), 36.
[120] Klein, Naomi, 36.
[121] Klein, Naomi, 26.
[122] Hayhoe, Katharine, *Saving us:* 38.
[123] Reidmiller, 200.

Figure 20a: Natural and Human Influences on Global Temperature[124]

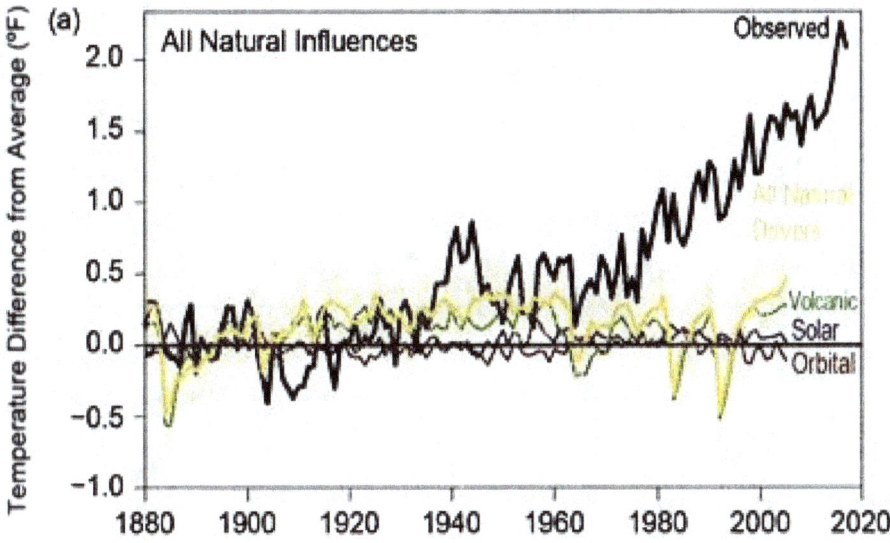

Figure 20a shows the "Observed" average temperature changes (black line) versus all other natural drivers combined (yellow line). Individual contributions are also plotted: sun cycles (purple line), changes in the Earth's orbit (brown line), and volcanic eruptions (green line). Natural causes cannot explain the long-term trend in average global surface temperatures from 1880 to 2017.

---

[124] Reidmiller, 200.

Figure 20b:

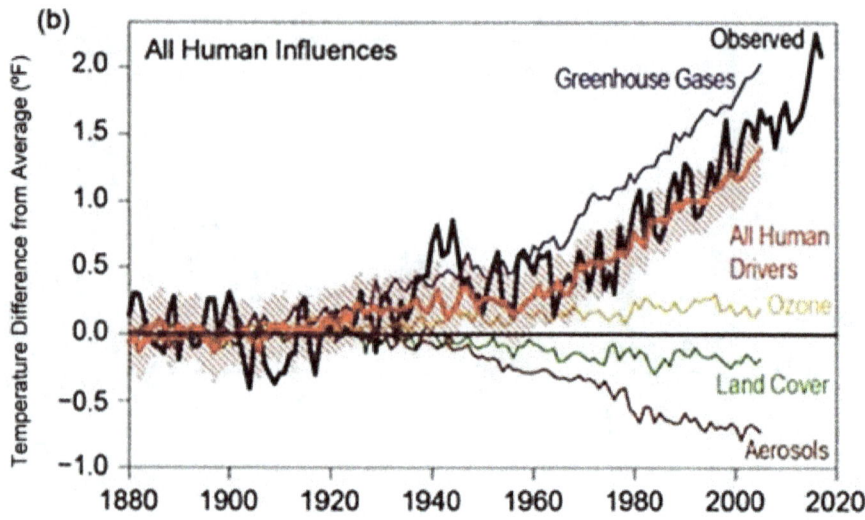

Figure 20b shows the average global temperature changes when only human causes are considered (red line). The human drivers include GHGs (purple line) and aerosols (brown line), land cover (green line), and ozone (yellow line). Aerosol emissions have a small net cooling effect. Human-produced GHGs (CO_2 and CH_4) are the dominant drivers of global temperature change.

Figure 20c:

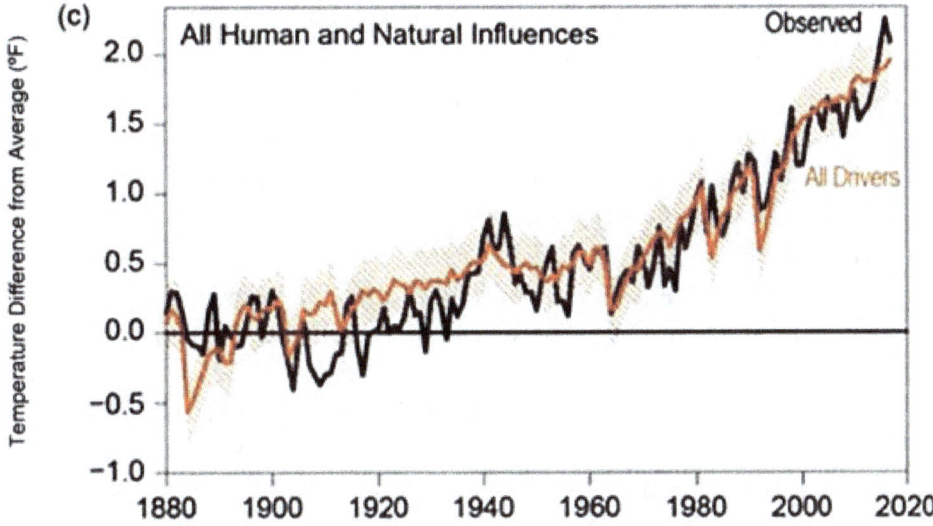

Figure 20c shows the average global temperature change predicted by computer models (orange line) when all natural and human drivers are included. The computer model result matches the observed temperature historical data quite well. In particular, the computer model is in excellent agreement from 1950 forward when human drivers were the greatest.

Some deny that climate change is real or happening at all. Others become cynical or fatalistic, thinking it is too late to take corrective actions. Others use pseudoscientific arguments to discredit the climate science claim that climate change is not human-caused. These arguments are called pseudoscience because they wrap something false around an actual scientific fact and present it as scientific. Those biased against climate science readily accept false, alternative causes for climate change, such as sun cycles, volcanic eruptions, orbital cycles, natural weather cycles, and wrong or made-up climate data. Climate scientists have thoroughly disproved these claims as the primary source of climate change.

2. <u>Climate action opponents</u>: Opposition to climate action has been effective, mainly due to the influence of corporations and politicians. They sound the false alarm that climate action will ruin the economy and destroy the standard of living of modern society. Many have lobbied to maintain the status quo on energy policy, convincing people that a transition to a clean energy

economy is financially impossible, effectively stalling progress in climate action.

Those supporting the continuation of the fossil-fuel-based energy economy have influential, well-aligned supporters in government and industry. They have known the effectiveness of playing on an individual's fears and sense of fair play to create personal inaction and collective opposition—climate action means personal sacrifice. By contrast, others reap the benefits at my expense.

Florida Governor Ron DeSantis turned down federal Inflation Reduction Act (IRA) climate action funding for his state.[125] If implemented, this program would provide funding and incentives for energy efficiency improvements and help low-income households make these improvements to lower energy bills. It is a step towards a more sustainable future, where we all benefit from reduced energy costs and a healthier environment. His move blocked climate change corrective actions that are proven effective and necessary to meet the 1.5 °C global warming goal; these programs would help improve people's lives and promote concern for the unfortunate—they are both moral and just. DeSantis, whose state has been hit by rapidly intensifying hurricanes, denies the impact of climate change on hurricane intensity despite scientific studies that show higher ocean temperatures lead to more severe storms. He signed legislation that prevents cities from converting to 100% renewable energy despite the example of Babcock, Florida, which runs on solar power and kept its electricity and water running during Hurricane Ian. He supports increased fossil fuel use and opposes rooftop solar installation in his state.[126] This approach puts his state on the path to increase, not reduce, GHG emissions.

DeSantis has garnered solid political support for his opposition to the so-called "liberal agenda." He has used the politics of division to play on personal fears that government mandates will impose an unfair burden on individuals, restrict lifestyle choices, and take away individual freedoms. The rejection of the liberal agenda and the climate denial message has taken root in Florida. DeSantis has successfully blocked climate action and supported the use of fossil fuels. Nevertheless, human-caused climate change has arrived in Florida and is intensifying. Since 1980, the frequency of storms and hurricanes that rapidly

---

[125] Dorn, Sara, "DeSantis Under Fire for Rejecting Millions in Home Energy Funding from Biden's Inflation Reduction Act," *Forbes*, (updated August 30, 2023) accessed August 31, 2023.

[126] McDowell, Tim, "Ron DeSantis's Climate Contradictions," *SEMAFOR*, (updated April 7, 2023), accessed August 31, 2023.

intensify has increased. Powerful hurricanes have brought high winds, heavy rains, and devastating storm surges. Lives have been lost, livelihoods ruined, and infrastructure destroyed. Costs to rebuild have risen into the hundreds of millions of dollars, insurance premiums have increased, and some insurance companies no longer offer hurricane coverage. Climate change does not care if you are a Democrat or a Republican. Climate change is real and human-caused, and its impacts are being felt today in our backyard.

3. Climate technology opponents: Still, others take a "more-with-less" approach, a stance that theologically and scientifically challenges technology-based climate action. This approach is based on the belief that technology alone cannot "solve" the climate change problem. It argues that the use of technology often creates more problems than it solves and cannot provide a sustainable and just solution to the climate crisis. However, it also offers hope, suggesting we can create a more sustainable and just world by using less and practicing good stewardship of our resources. Later in the book, we will explore the arguments of those who support a degrowth approach, which rejects technology solutions, promotes a return to a pre-industrial age, and intentionally reduces the standard of living to benefit life-giving ecosystems and human flourishing.

4. Climate science and technology proponents: A third approach accepts climate change as real and human-caused, does not give up on energy technology, and embraces an accelerated transition to net-zero energy production for transportation, heating, and industry. This argument claims that using net-zero technologies combined with more efficient energy use results in much less harm to life-giving ecosystems than the current inefficient, carbon-emitting, fossil-fuel-based approach. This transition reduces harm and offers the potential for a more sustainable and just world, encouraging and motivating us to take action. It can be accomplished in a just manner while maintaining a moral and sustainable standard of living.

Historically, opposition to climate action has been effective. However, there is a growing acceptance of the scientific consensus that climate change is real and human-caused. This increasing public acceptance of climate science has prompted some governments and businesses to take steps to reduce fossil fuel consumption. The power of public opinion in changing the climate change narrative is evident in a recent study by Andre et al. published in *Nature Climate Change*, which has "revealed widespread support for climate action. Notably, 69% of the global population is willing to contribute 1% of their income, 86% endorse pro-climate social norms, and 89% demand intensified political action."[127] This shift in public opinion is a significant step forward in the fight against climate change. The influence of public opinion can make us feel empowered and influential in the fight against climate change.

Andre et al. demonstrate that challenges to implementing effective climate policies are due to a false perception that individuals are unwilling to take climate action. Their work identifies this as the "perception gap." Another challenge to climate action identified in the book is individuals showing "conditionally cooperative behavior." Raising awareness of broad support for climate action and promoting a unified response to climate change significantly improves people's "willingness to cooperate (WTC)." We need to raise awareness of the science-based root causes of climate change to debunk false ideas that undergird climate denialism. We must discuss ways people can individually take climate action to promote new, ecosystem-friendly behavioral norms. We also need to advocate for the development and implementation of climate policies through political action.

5. Climate change and the religious divide: Many Christians are driven by political partisanship, and culturally informed ideological viewpoints are often used to defend their religious beliefs. Religious perspectives justify the rejection of climate science and its causes and consequences. Dr. Katharine Hayhoe, an evangelical Christian climate scientist, points out that due to the "influence of culture and religion," more than two-thirds of white evangelical Christians in the United States maintain that climate change is not human-caused. Many fundamentalist Christians dismiss the threat of climate change, oppose climate action, and argue for economic growth based on the belief that God is in control of the earth and only God can save it. According to the theologian Joseph R. Wiebe, some Christians believe that environmental outcomes are "not humanly determined and therefore not up for debate."

The following discussion compares and contrasts two biblically based approaches to using the earth's resources: the "dominion" approach from Genesis 1:28 versus the "till and keep" approach from Genesis 2:15.

The dominion approach asserts that God grants humans dominion over the earth, implying that the earth is for human use. God's care for the earth

---

[127] Andre, Peter, et al., "Globally Representative Evidence on the Actual and Perceived Support for Climate Action," *Nature Clime Change*, 14, (March 2024): 253-259.

surpasses anything humans can do to harm it. Humans are to worship God, not creation, rely on God's provision, and trust God for salvation. An uncritical acceptance of the "dominion" approach may lead to a form of spiritual bypass that neglects human responsibility and the seriousness of the situation.

Following the dominion approach without proper evaluation may justify poor stewardship and lead to irresponsible use of the earth's resources. A case in point is whale hunting. For thousands of years, indigenous groups have hunted whales sustainably. Whale meat and blubber served as food sources, while other products were vital for survival. Little went to waste. The number of whales harvested each year could be naturally replenished, so extinction was not a threat. In the 1800s, Western Europeans developed new harpoons and faster whaling ships to hunt the highly valued blue whales. By 1900, blue whales were nearly extinct; 99% had been killed primarily for their blubber. Blubber became a valuable source of lamp oil, machine lubricant, and was used to make perfumes. Since the hunting of blue whales was outlawed in the 1960s, their numbers have increased.

Why do we need blue whales? Blue whales are part of the ocean ecosystem. They eat krill and produce excrement that feeds plankton, which all sea life, including krill, depends on. Plankton removes CO_2 from the atmosphere and releases O_2. Removing CO_2 from the atmosphere and releasing O_2 benefits the biosphere. Blue whales are an essential part of the ecosystem. Hunting whales sustainably for community survival exemplifies responsible dominion. In contrast, hunting blue whales for profit to the point of extinction harms the ecosystem God created and called good, and is not responsible dominion.

Many Christians who practice a "till and keep" approach believe God commands humans to be good stewards of what God provides. They accept scientific findings on climate change and advocate for sustainable and just climate action. Informed action produces hope and provides purpose to one's life.

"Till and keep" Christians believe that their actions make a difference. They work as a community to care for the earth's life-giving ecosystems, make sustainable and just use of its resources, and practice spirituality that seeks to reconcile the alienation of broken relationships with God, self, others, and the earth. This commitment stems from the belief that God created the earth, called it good, and entrusted us with its use and care. The "till and keep" approach fosters a balance that encourages responsible use and sustainable care of God's creation.

6. Climate action advocates: Many environmentalists take a secular approach to the climate crisis. These climate change advocates accept climate science and know that the time to act is now. They take personal and collective action on climate change.

People in the secular community who accept climate scientists' research findings and are concerned about the environment align with Christians who follow the "till and keep" approach and practice creation care. They form a climate change advocacy group that promotes viable alternatives to fossil fuels. Their voices would address those claims that oppose or cast doubt on the way forward. Forecasts of economic ruin may be countered with the feasibility of successfully transitioning from fossil fuels to renewable, clean energy. Making this case before a deeply entrenched opposition is a daunting task. The transition from fossil fuels to renewable energy must occur to save the earth from eventual climate catastrophes. Environmentalists and Christian creation care advocates need to speak out about the clean energy transition and work to inform a misinformed opposition by educating supporters and opponents on the reality of the climate crisis and the way forward. They must use consumer purchasing power to promote a sustainable and fair transition to a clean energy economy. They must contact politicians and advocate for policy changes and programs that support and accelerate the clean energy transition.

Chapter 5. Climate Action

This next section discusses how the Earth's ecosystems are interconnected and how human activity causes environmental harm beyond climate change. This section will describe ways to discover, develop, and deploy meaningful, timely, and constructive climate actions that lead to long-term, just, and sustainable environmental outcomes. We will see how these approaches must be implemented globally, using the best knowledge and employing an interdisciplinary network of corporation leaders, government officials, environmentalists, consumers, and citizen groups. It will require the skill and know-how of scientists, engineers, biologists, architects, community planners, economists, and bankers to design and implement plans to achieve climate change goals.

This section points to the importance of informing, supporting, and electing politicians who promote a sustainable transition to a clean energy economy. Who we vote for and what we do make a difference. We must ask: Who are the leaders who give climate action the exposure and priority it demands? What climate action programs do we support? What climate actions are we personally willing to take?

Eco-overshoot to Sustainability

In her chapter from *Drawdown*, Andrea Wulf writes about the work of Alexander von Humboldt, a 19th-century pioneer in biogeography. In the early 1800s, Humboldt revealed that "nature is intricately connected in ways that surpass human knowledge... [Humboldt documented how] living systems were vulnerable to disturbance by human beings."[128] He examined the relationship between humans and nature. Humboldt's work identified ways humans impacted the climate through deforestation, the draining of lakes, the drying of marshes, and the polluting of streams with steam and gas emissions from industrial equipment.

Up to now, this book has focused on humans' impact on the climate through GHG emissions. However, human activity is impacting the Earth's ecosystems in multiple ways. The disturbance of one ecosystem spills over into

---

[128] Wulf, Andrea, "Alexander von Humbolt," *Drawdown: The Most Comprehensive Plan Ever Proposed to Reverse Global Warming*, editor Hawken, Paul, (Penguin: New York, 2017), 24-25.

disturbances in other parts of the environment. This means that working to achieve environmental sustainability goes beyond addressing climate change alone. As the interrelatedness of the Earth's ecosystems has come into view, Johan Rockstrom and a group of 28 scientists identified an interconnected framework of ecosystem processes that regulate the Earth's environment, called the nine planetary boundaries.[129] These scientists have quantified nine planetary boundary limits "within which humanity can continue to develop and thrive for generations to come."[130]

Their work quantifies human activity's impact on the Earth's environmental processes. The environment's natural processes can repair and restore the damage caused by human activity. Ecological sustainability is maintained so long as the environment's capacity to repair itself exceeds the damage done by human activity. Eco-overshoot occurs when human activity exceeds the environmental system's capacity to repair waste production damage and restore resource extraction loss. Rockstrom's research indicates that six planetary systems are in eco-overshoot. Transitioning from unsustainable eco-overshoots to sustainable eco-efficiency requires human effort on a global scale.

A recent study by Richardson et al. documents the Impact of human activity on the planetary boundaries.[131] The data in Table 5 shows that human activity has exceeded the safe operation limits in six of the nine boundaries (highlighted in red). Figure 21 illustrates the same data.

---

[129] Rockstrom, Johan, et al., "Planetary Boundaries: Exploring the Safe Operating Space for Humanity," Ecology & Society," vol 14(2), 2009.

[130] Stockholm Resilience Centre, "Planetary Boundaries," Stockholm University, (published September 2023), accessed October 17, 2023.

[131] Richardson, Katherine, et al., "Earth Beyond Six of Nine Planetary Boundaries," Science Advances, (published September 13, 2023), accessed October 18, 2023.

Table 5: Current Status for the Planetary Boundaries[132] (adapted from Richardson et al)

Earth System Process	Control Variable(s)	Planetary Boundary	Preindustrial Value	Current Value
Climate Change	CO2 level (ppm)	350	280	**417**
	Atmospheric heat flux (W/m^2)	1.0	0.0	**2.9**
Stratosphere Ozone Loss	Ozone thickness (DU)	276	290	285
Atmospheric Aerosols	Aerosol Depth (AOD)	0.100	0.030	0.076
Ocean Acidification	Acidity of ocean water (-0.1 pH = +26% acidity)	8.07	8.20	8.10
Fertilizer Flows	Phosphates (Ktons/year)	11	1	**18**
	Nitrogen (Ktons/year)	62	0	**190**
Freshwater Availability	Blue Water (groundwater consumption)	10	9	**18**
	Green Water (rainwater consumption)	11		**16**
Novel Entities Pollution	Toxic Synthetic Chemicals Released to the Environment	0	0	**?**
Land System Change	Forest Land Conservation (% remaining)	75	100	**60**
Biosphere Integrity	Genetic Diversity (extinctions/million species/yr)	10	1	**>100**
	Ecosystem Functioning (% Biomass consumed)	10	2	**30**

---

[132] Richardson, Katherine, et al., "Earth Beyond Six of Nine Planetary Boundaries," *Science Advances*, (published September 13, 2023), accessed October 18, 2023.

Figure 21: Current Status for the Planetary Boundaries[133] (from Richardson et al)

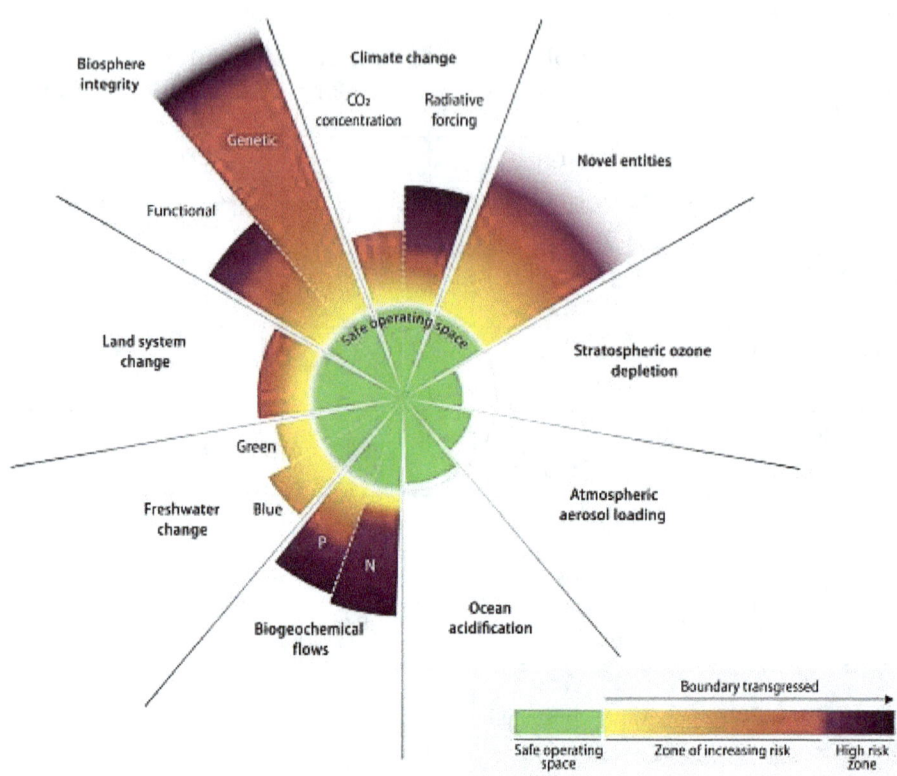

The green zone represents the range in which humans can safely operate, continue to develop, and thrive for future generations.

This figure indicates that 6 of the nine planetary boundaries have been transgressed.

Sustainable use of the Earth's resources must consider the processes that human activities affect.

---

[133] Richardson, Katherine, et al., "Earth Beyond Six of Nine Planetary Boundaries," *Science Advances*, (published September 13, 2023), accessed October 18, 2023.

The science behind the planetary boundaries is complex. The table shown above is a very simplified summary of this work. According to Richardson, these boundaries are not tipping points but warning signs. "It is like blood pressure. If your blood pressure is 120 over 80, it is not a guarantee that you will have a heart attack, but it raises the risk, and therefore, we do what we can to bring it down."[134] The global impact of climate change interacts with other planetary ecosystems, putting them under more significant stress. Increased CO_2 levels in the atmosphere raise ocean acidification, which affects marine life. Climate change affects weather patterns that affect freshwater availability, which impacts genetic diversity and ecosystem functioning. Increases in fertilizer use and runoff expand GHGs and affect aquatic ecosystems. Land management change, primarily the loss of rainforest due to agricultural expansion, affects the climate and threatens genetic diversity and ecosystem functioning.

A case in point is that from 2018 to 2021, about 10 billion snow crabs disappeared from the waters off the coast of Alaska.[135] The snow crab population declined so much that the Alaska Department of Fish and Game canceled the snow crab season. Recent findings published in *Science* reveal that snow crabs experienced a mass starvation event caused by a marine heat wave.[136] Due to climate change, the Bering Sea waters have warmed faster than the rest of the planet. Snow crabs seek colder water in the summer to survive, but the Bering Sea heat wave significantly decreased the amount of cold water available. This caused the crabs to concentrate in a smaller space, and higher water temperatures increased the crabs' metabolic rate. The increased number of crabs in a smaller pool with greater caloric needs resulted in mass starvation. This disruption of the Bering Sea ecosystem, triggered by climate change, has significantly affected the Alaskan fishing industry.

This example shows the interconnectedness of ecosystems—Bering Sea warming resulted in a mass die-off of snow crabs. It shows how human activity impacted the biosphere—the burning of fossil fuels increased GHGs in the atmosphere, which caused the Bering Sea to warm.

---

[134] Bartels, Meghan, "Humans Have Crossed 6 of 9 'Planetary Boundaries'," Scientific American, (posted September 13, 2023) accessed October 17, 2023.

[135] Bush, Evan, "In Search of 10 billion Missing Snow Crabs, Scientists Eye Marine Heat Waves," *NBC News*, (published October 20, 2023) accessed October 21, 2023.

[136] Szuwalski, Cody, S., et al., "The Collapse of Eastern Bering Sea Snow Crab," Science, vol 382 (6668), (published October 19, 2023), accessed October 20, 2023.

The proper functioning of ecosystems is crucial for the long-term health of our environment. Climate change has stressed ecosystems, which has led to reduced robustness and resiliency. Ecosystem functioning involves the ability to break down biosphere contamination from human-generated wastes and the ability to regenerate biosphere resources that human activity has depleted. As human activity exceeds the safe operating limits of the planetary boundaries, the Earth's capacity to restore and regenerate diminishes, leading to the loss of the biosphere's functional integrity and resiliency. This means the Earth becomes increasingly contaminated with human waste and has fewer resources (such as plants, trees, insects, marine life, birds, and animals) for human use. Humans must learn to live within the nine planetary boundaries to continue developing and thriving for generations.

The Earth's environmental system is reaching the limits of its capacity to sustain human economic growth. The authors of *Limits to Growth* state, "We know absolutely that physical growth on this planet will stop."[137] In its current waste production and resource consumption form, the human enterprise cannot flourish on a planet with limited resources. Eventually, human activity will exceed the ecosystem's assimilating and regenerative capacities. The authors ask, "What can be done to create a human economy that provides sufficiently for all?" And "Are the current policies leading to a sustainable future or to collapse?"[138]

Two factors weigh heavily on the future outcome of the human enterprise: (1) how much capacity does the Earth have to absorb and process waste from human activity, and (2) what are the consumption limits of the Earth's natural resources? The answers to these questions define the limits to growth and should direct policies that allow humans to flourish sustainably for generations to come.

The authors of *Limits to Growth* assert that humanity's "ecological footprint has exceeded the carrying capacity of the earth."[139] This means their assessment indicates that human activity is already in a state of eco-overshoot (i.e., producing waste in a greater amount and at a higher rate than the Earth can process and extracting more resources at a higher rate than the Earth can

---

[137] Meadows, Donella, et al., *Limits to Growth: The 30-Year Update*, (White River Junction: Chelsea Green Publishing, 2004), xi.

[138] Meadows, Donella, et al., xii.

[139] Meadows, Donella, et al, xix.

replenish), which is unsustainable over the long term. They maintain that higher eco-efficiency (i.e., producing less waste and making better use of resources) is needed to move out of the current state of eco-overshoot.

Humans must accept their role and responsibilities in the deterioration and exploitation of the Earth's biosphere. They must move back from the precipice of the eco-overshoot into sustainable eco-efficiency. According to the authors of *Limits to Growth*, human enterprises face inevitable and "sudden" collapse unless they make the necessary policy and behavioral changes.[140]

Is Technology Part of a Just Solution?

Some see the technology-based, clean-energy transition as fundamentally flawed because it cannot be accomplished justly and sustainably. Some support a degrowth approach that shuns technology solutions, promotes a movement towards a pre-industrial age, and intentionally lowers the standard of living to benefit life-giving ecosystems and human flourishing.

Some community planners, seeking meaningful and timely climate action and wanting just and sustainable human activity, hold to the premise that the eco-overshoot crisis cannot be solved by using human technology. They claim the Earth is in an extreme state of eco-overshoot; humans are "exploiting ecosystems beyond their regenerative assimilative capacities."[141] Their assessment of renewable energy technology indicates that metals mining and waste production "entail egregious social injustices and significant ecological degradation."[142] They maintain that the only way out is through managed descent to a lower standard of living and population level.

This perspective challenges the notion that the Green New Deal (GND) can provide an ecologically sustainable solution to the climate crisis and that the clean energy transition (CET) is riddled with social injustices and causes ecological harm. This analysis concludes that there are better ways forward than the GND and CET approaches. Instead, it proposes a planned reduction in the standard of living and a cooperative decline in the population as alternatives. This approach, it is argued, will reduce energy demand, waste production, and

---

[140] Meadows, Donella, et al, xxv.

[141] Seibert, Megan K., Rees, William E., "Through the Eye of a Needle: An Eco-Heterodox Perspective on the Renewable Energy Transition," *Energies* 14, no. 15: 4508, p. 2.

[142] Seibert, Megan K., Rees, William E., p. 3.

84

consumption of natural resources to a sustainable level and pave the way for a more sustainable relationship with our environment.

As Siebert and Rees assert, "The only viable response to eco-overshoot is a managed contraction of the human enterprise…"[143] Contraction is the only way to arrive at a place where humans can safely operate within the bounds of the Earth's ecosystems and thrive for generations to come. Siebert and Rees' approach involves fewer people than today, consuming less of the Earth's resources. They say, "The world must abandon neoliberal capitalism's material growth imperative and face head-on that material life after fossil fuels will closely resemble life before fossil fuels."[144] What seems to be implied here is that humans must return to the pre-industrial standard of living.

The success of this approach relies "upon unprecedented levels of global cooperation…" and the exertion of "unrelenting pressure" on "ruling elites" such that they "have no choice but to capitulate to specific, well-thought-out demands."[145]

While the concept of managed descent has merit, Siebert's and Rees's approach is idealistic, and their method is untenable. The demands to lower the standard of living and reduce the population will be rejected by the powerful elites and strongly resisted by the populists. Nonetheless, this work cannot be dismissed outright. Indeed, there is a need to define a sustainable and just standard of living and to work toward achieving that end.

Another point raised in the Siebert-Rees article is the comparative environmental impact of coal, oil, and gas extraction versus mining for metals. Power plants and engines burn fossil fuels. Windmills and electric motors require metals to make magnets and batteries. The clean energy transition would shift from fossil-fuel-based electricity and transportation to a metals-based energy system. The extraction of fossil fuels and mining minerals each has its unique environmental footprint. According to an MIT environmental scientist, Scott Odell, comparing the two technologies "Isn't straightforward… making an apples-to-apples comparison is challenging."[146]

---

[143] Seibert, Megan K., Rees, William E., p. 3.
[144] Seibert, Megan K., Rees, William E., p. 12.
[145] Seibert, Megan K., Rees, William E., p. 15.
[146] Ferreira, Fernanda, and Odell, Scott, "How Does the Environmental Impact of Mining for Clean Energy Metals Compare to mining for Coal, Oil, and Gas?" *MIT Climate Portal*, (published May 8, 2023) accessed October 28, 2023.

As the clean energy transition moves forward, the demand for fossil fuels will drop, and the demand for metals will increase. Fossil fuels are burned; they can only be used once. The metals used to make windmill generators and EV batteries can be recycled and used again. The burning of fossil fuels produces CO_2. The operation of windmills and EVs does not emit CO_2. Odell cites the work of the International Energy Agency (IEA), "The emissions created by extracting minerals from the ground are tiny compared to those created by burning fossil fuels. A 2020 report from the IEA found that for every gigawatt of a clean energy technology installed, millions of tons of CO2 emissions can be avoided." [147, 148]

According to Odell, "Given the urgent threat of climate change, the clean energy transition is necessary."[149] The sustainability of clean energy mining rests on three changes: (1) Reduce the need for cars by increasing public transportation. (2) Reuse metals and materials in windmills and EVs to reduce the extraction of raw materials. (3) Raise mining industry standards to ensure mining is done safely, in an environmentally sustainable and socially responsible way.

Do We Have the Needed Technology, and Can We Get to Net-Zero Emissions in Time?

Some support accelerating the clean energy transition through sustainable and just technology solutions. The book *Drawdown*, edited by Paul Hawken, presents a comprehensive list of "climate solutions that have the greatest potential to reduce emissions or sequester carbon from the atmosphere."[150] The actions on the list consider the availability of the solution, its effectiveness, and the cost. *Drawdown* experts evaluated the inputs and performed the calculations. They developed a model that predicted the relative effectiveness of each solution. The results of the calculations were reviewed and validated by a 120-person board of experts. The results identify viable pathways to meet the Paris Agreement and COP 26 goals. Implementing these climate actions

---

[147] Ferreira, Fernanda, and Odell, Scott.

[148] Cozzi, Laura, "Sustainable Recovery," *International Energy Agency*," (revised July 2020), accessed October 28, 2023.

[149] Ferreira, Fernanda, and Odell, Scott.

[150] Hawken, Paul, editor, *Drawdown: The Most Comprehensive Plan Ever Proposed to Reverse Global Warming*, editor Hawken, Paul, (Penguin: New York, 2017), x.

produces hope for the future. Hawken states, "This is not a liberal agenda, nor a conservative one. This is a human agenda."[151]

The transition to a clean energy economy must meet GHG emission goals promptly. This will require a sustained political effort. People must hold their leaders accountable for meeting the Paris Agreement limit of 1.5°C by the end of the 21st century and the COP26 goal of net-zero emissions by 2050. We must answer whether we have the technology to limit temperature rise to 1.5 °C to achieve net-zero emissions in time.

Jonathan Foley, Executive Director of *Project Drawdown*, in the video "Climate Solutions 101: Putting it All Together," [152] categorizes the *currently available* technologies and climate actions into three categories: 1) Sources: reduce emission sources from electricity generation, food/agriculture/land use, industry/manufacturing, buildings/construction, and transportation/shipping, 2) Sinks: support natural CO_2 sinks to remove CO_2 from the atmosphere, and 3) Society: improve society through health and education to provide equity for all. Figure 22a summarizes these approaches. The authors of *Drawdown* provide a detailed analysis of the impact of climate action on CO_2 and each climate action's economic merits.[153]

---

[151] Hawkins, Paul, xi.

[152] Foley, Jonathan, "Climate Solutions 101: Putting it All Together," *Project Drawdown*, (published), accessed October 20, 2023.

[153] Hawken, Paul, editor, *Drawdown: The Most Comprehensive Plan Ever Proposed to Reverse Global Warming*, editor Hawken, Paul, (Penguin: New York, 2017).

Figure 22a: Project Drawdown Climate Actions[154]

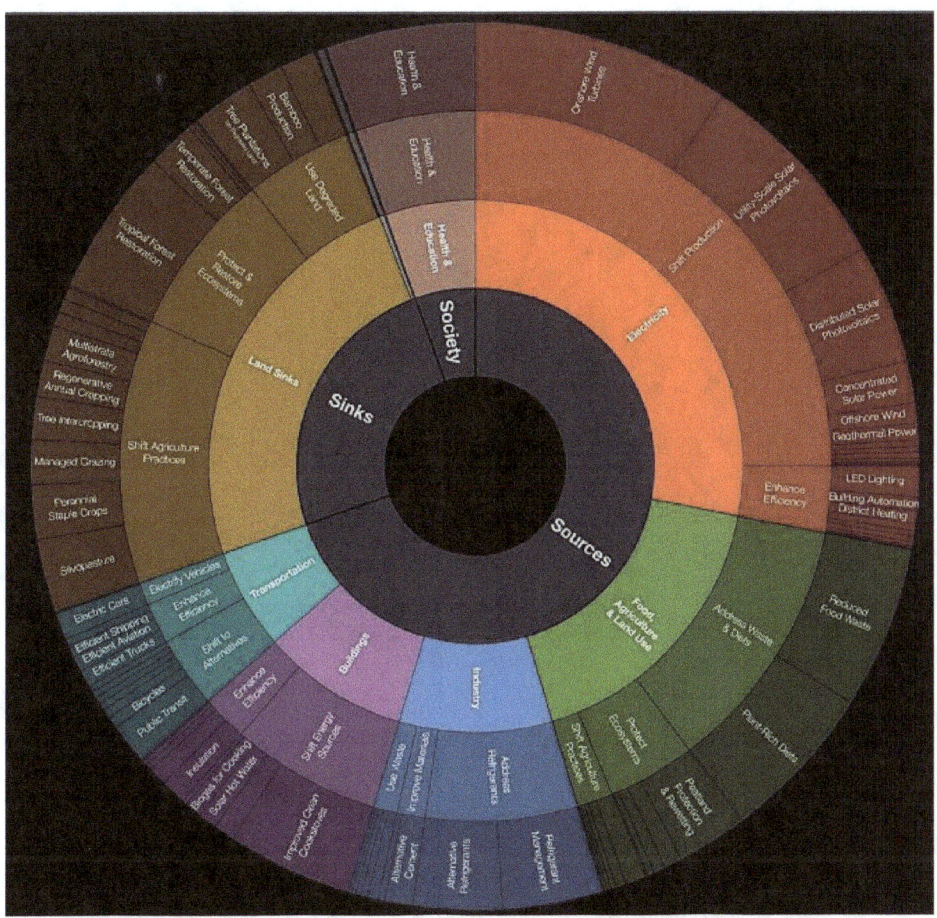

In the video, Foley quantifies the impact of climate actions on atmospheric CO_2 concentrations and global temperatures. Modeling results show the relative effects of three climate action scenarios in Figures 22b and 22c. The Earth has an atmospheric CO_2 level of about 420 ppm and a global temperature increase of 1°C above preindustrial levels. The baseline scenario (red curve), which implements only those climate actions already in place, reaches 725 ppm CO_2 and a 2.5°C global temperature increase by 2060. Scenario 1 (yellow curve), which implements currently promised climate actions, reaches 550 ppm CO_2

---

[154] Foley, Jonathan, "Climate Solutions 101: Putting it All Together," *Project Drawdown*, (published), accessed October 20, 2023.

88

and a 2°C temperature increase by 2060. Scenario 2 (green curve), which implements all of the drawdown climate actions, reaches a *maximum* CO_2 concentration of about 475 ppm and a global temperature increase of 1.5°C by 2050.

Figure 22b: Impact of Climate Actions on Atmospheric CO_2 Levels[155]

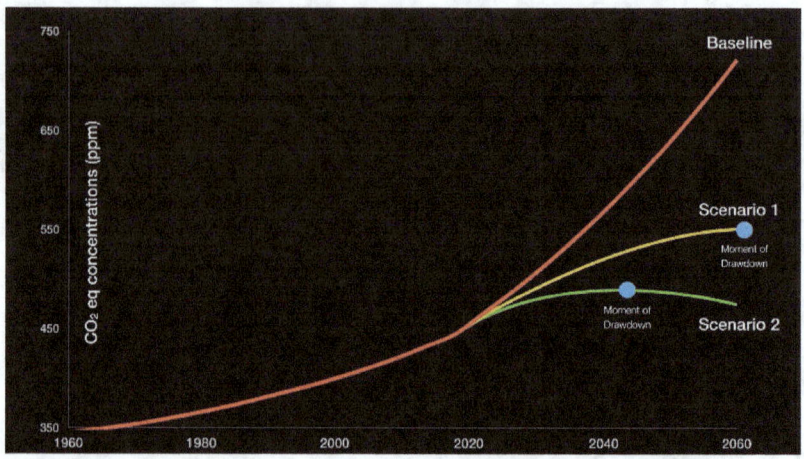

Figure 22c: Impact of Climate Actions on Global Temperatures[156]

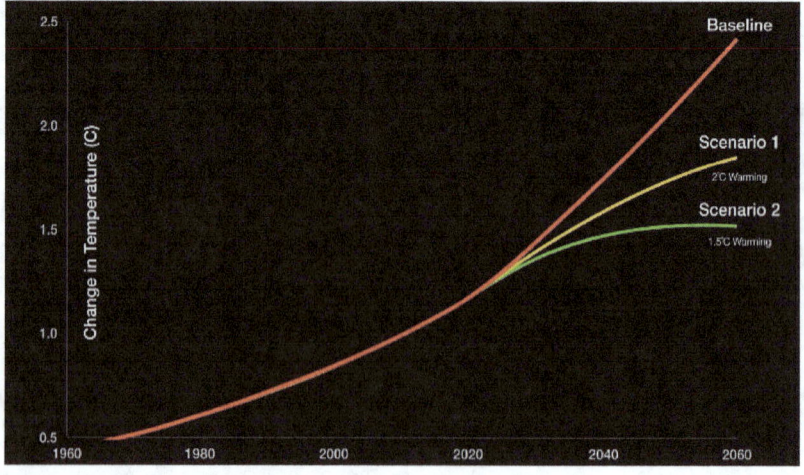

[155] Foley, Jonathan, "Climate Solutions 101: Putting it All Together," *Project Drawdown*, (published), accessed October 20, 2023
[156] Foley, Jonathan, "Climate Solutions 101: Putting it All Together," *Project Drawdown*, (published), accessed October 20, 2023

The Drawdown Project helps to answer the question, "Does the clean energy transition make sense for humanity?" Humans can flourish through the responsible use of the Earth's resources, the wise care of life-giving ecosystems, and the just treatment of people. Humans have the technology to achieve the COP26 net-zero emissions goal by 2050 and limit the post-industrial global temperature rise to the Paris goal of 1.5 °C.

Table 6 shows a partial list of the Drawdown Project climate action benefits: CO_2 reductions and cost savings over 30 years.

Table 6: Partial List of Drawdown Climate Actions: CO_2 Impacts and Cost & Savings[157]

Category	Sector	Climate Action (from *Drawdown* pp 224-225)	CO2 Reduced (GT)	Net Cost ($B)	Net Saved ($B)
Sources	Energy	Wind (onshore + offshore)	99.8	1,771	8,188
		Solar (rooftop + solar farms)	61.5	373	8,482
	Transport	EVs, Cars, Trucks, Mass Transit	27.6	14,091	16,591
		Ships, Planes	12.9	1,578	3,612
	Buildings	"District" Heating/Cooling	9.4	457	3,544
		Insulation	8.3	3,656	2,513
		Heat Pumps/Smart Thermostats	7.8	45	2,187
		LED Lighting (home + business)	12.8	119	2,820
	Materials	Refrigerants (See Note 1)	89.7		-903
		Alternate Cement Production	6.7	-274	
		Home Water Efficiency	4.6	72	1,800
		Recycling (home + industrial)	6.4	1,307	142
	Food, Agriculture	Reduce Food Waste	70.5		
		Plant Rich Diet	66.1		
		Farmland Restoration	14.8	72	1,342
Sinks	Land Use	Tropical Forest	61.2		
		Temperate Forests	22.6		
		Peatlands	21.6		
		Afforestation (See Note 2)	18.1	47	1,015
Society	Women, Girls	Education	59.6		
		Family Planning	59.6		
Totals			741.6	23,314	51,333

---

[157] Hawken, Paul, editor, *Drawdown: The Most Comprehensive Plan Ever Proposed to Reverse Global Warming*, editor Hawken, Paul, (Penguin: New York, 2017), 220-225.

Note 1: Refrigerants: Although found at low levels, chlorofluorocarbons (CFCs) and hydrofluorocarbons (HFCs) can trap heat greater than 1,000 times the amount of heat CO_2 traps. These refrigerants leak into the atmosphere. Note 2: Afforestation is land conversion by planting trees for commercial harvest.

Each climate action does one or more of the following: reduces energy use through material reduction and efficiency and productivity increases; replaces fossil fuel energy resources with renewable energy systems; sequesters carbon by growing trees, plants, and kelp. The research models use accurate, reliable, current data, and model parameters are checked for significance to increase confidence in the results. The models predict CO_2 reductions, costs to install and operate technologies, and savings over 30 years. The modeling shows that these climate actions have ecological, financial, and social benefits and that humans can achieve the goal of net-zero GHG emissions with known technologies.

The model's cost analysis shows a substantial return on investment. This analysis does not include the cost of climate change we avoid. There will be a very high catastrophic impact on society if we allow the planet to warm beyond 2°C. Hundreds of millions of lives will be affected, and severe economic consequences will be incurred.

These climate actions provide jobs that create economic security, improve people's health, reduce hunger, and help increase resiliency. These actions benefit human flourishing for future generations. There is nothing to stop us from taking climate action today. We have enough climate solutions to stop climate change. We must advocate for government and business action.

These climate actions depend upon humans working together with the common goal of achieving personal well-being and the well-being of future generations. There are things we all can do! Everyone has a role and a responsibility to take climate action. What climate actions can you take, and what actions will you take? Together, we make a difference that matters!

Part III

Creation Care, Climate Justice,

& Ecospirituality

Chapter 6. A Theology of Creation Care – Section 1

Human theologies, religious beliefs, and practices influence our values and priorities. Traditional theology and religion encourage nurturing our relationships with God, self, others, and all creation. These relationships are the basis for stewardship of the earth's life-giving resources. In this section, we will discuss three main approaches: a theology of Creation Care, which exposes the four-fold alienation and reveals the path to reconciliation with God's creation; Climate Justice, which emphasizes the right treatment of our neighbors in the shift to sustainable living; and Ecospirituality, which involves confessing our past mistakes, finding reconciliation, and being grateful for God's restorative mercy.

Prophetic Exhortation

How do humans develop the economic and political will to make the changes that could avert a climate catastrophe? Many claim that shifting away from the Western metanarrative of progress by reducing consumption and letting go of some lifestyle pleasures is not a workable way to avert a climate crisis. These actions would harm economic prosperity and require burdensome personal sacrifice. Humans want to protect the economy, which provides for basic needs and material pleasure, and they want to maintain their lifestyle even if it means the whole of creation will suffer. The reality of the climate crisis exposes the consequences of unsustainable growth. The mismanagement of resources and the physical destruction of the environment lead to a climate catastrophe. The amount of human suffering produced by climate change creates unjust suffering and, therefore, is a moral crisis.

In Mark 9:43, Jesus warns, "If your hand causes you to sin, cut it off. It is better for you to enter life maimed than with two hands to go into hell [*geennan*], where the fire never goes out."[158] *Geennan* [in Hebrew called: *gehinnom*] refers to the place near Jerusalem where children were sacrificed as burnt offerings to Moloch, a bull-headed idol (Je 7:31). Later, *geennan* became a trash dump for the city (2 Ki 23:10).

Perhaps in this context, Jesus speaks to a moral crisis and provides a prophetic assessment. Jesus warns about the causes of immoral behavior,

---

[158] Ryrie, Charles Caldwell, *The Ryrie Study Bible: Expanded Edition, New International Version*, (Chicago: Moody Press, 1994), 1540.

harming others, sin, and the consequences of bearing guilt for harm done. Suppose we lack the moral will to "cut off" or let go of what causes us to sin— e.g., the irresponsible destruction of life-giving ecosystems for unjustifiable material pleasure and love of mammon (a false object of worship that produces greed and covetousness). In that case, the desire for unsustainable material pleasure will eventually do irreversible harm to many of the Earth's life-giving ecosystems. Our self-indulgent prosperity will make our children and grandchildren victims of ecosystem collapse. The choice belongs to us. Should we cling to what causes us to sin or let go and be saved?

Hayhoe presents a point of view that recognizes human dominion *and* responsibility.[159] In Genesis 1:26, humans are made in the image of God and are to rule (have dominion) over every living creature on Earth (1:28). Humans are to rule the way God does, walking among the subjects (3:8) and caring for them. Created in God's image, humans are to rule over living creatures in a loving, caring relationship. In 2:15, God places humans in the Garden and gives them the responsibility to cultivate and preserve it. God also gives humans responsibility for every living thing on the earth. Dominion over the earth does not authorize humans to exploit the earth. God maintains a relationship with all creation, human and nonhuman. Each part of creation, living and nonliving, received God's blessing as "good" (Gen 1). Human exploitation of natural resources and destruction of the earth's life-giving ecosystems "is sinful because it is a violation of the God-creation relationship."[160]

The God-given blessing of dominion over the earth and the responsibility to care for the earth call for humans to stop unsustainable exploitation of the earth's resources and ecosystem destruction. The burning of fossil fuels, leading to a climate catastrophe, violates God's command to care for the earth responsibly.

Revelation 11:18 warns, "The time has come for judging the dead… and for destroying those who destroy the earth." This passage follows the sounding of the seventh trumpet (11:15), which announces the advent of the Messiah's "reign forever and ever" (11:16). The time for the final judgment "has come" (11:18). God's wrath will come against the wicked and immoral. God will "destroy" those whose sinful and corrupt behavior destroyed the earth. The verb

---

[159] Hayhoe, Katherine, *How to Save a Planet* Podcast.
[160] Dula, Peter, "Anabaptist Environmental Ethics: A Review Essay," *Mennonite Quarterly Review* 94, no. 1 (January 2020): 10.

"to destroy" [*diaphtheiró*] conveys the meaning of to spoil, completely deteriorate, fully decay, waste, ruin in a moral sense, or utterly corrupt.[161] The wicked and sinful ones whose corrupt, immoral actions laid waste to the earth and spoiled its resources will receive utter destruction as their reward. God's wrath is directed against those who destroy the earth's life-giving ecosystems and, as a consequence, cause great injustice. Climate change has physical and moral consequences.

In 2021, Pope Francis addressed the Scottish Catholic community during the COP26 climate conference. The Pope hoped the summit attendees would "meet this grave challenge with concrete decisions inspired by responsibility toward present and future generations." The Pope went on to say, "Time is running out; this occasion must not be wasted, lest we have to face God's judgment for our failure to be faithful stewards of the world he has entrusted to our care."[162] According to Huq, a leading climate scientist and biologist from Bangladesh, "Climate change is the greatest weapon of mass destruction of our times. Unless we… recognize this fact and do something about it, we are guilty of crimes against humanity."[163] The projected famines, forced migrations, and conflicts resulting from climate catastrophe represent our time's most significant moral issues.

The Separation of Heaven from Earth

Biblical scholar Howard A. Snyder informs us that, for many Christians, heaven is "the kingdom of ultimate personal fulfillment.". Richard Sterns (President of World Vision U.S.) "decries the limited view of the kingdom found in many evangelical churches."[164]

> *"Focusing almost exclusively on the afterlife reduces the importance of what God expects of us in this life. The kingdom of God in us was intended to change and challenge everything in our fallen world in the here and now. It was not meant to be a way to depart from the world but rather the means to*

---

[161] Arndt, William, F. and Gingrich, Wilbur, F., *A Greek English Lexicon of the New Testament and Other Early Christian Literature*, (Grand Rapids: Zondervan Publishing House, 1963), 180.

[162] Arocho-Esteves, Juno, "Failure to protect creation will mean facing 'God's judgment,' Pope Francis says," *EarthBeat*, (Nov. 11, 2021), accessed Dec. 9, 2021.

[163] Snyder, Howard A., 156.

[164] Snyder, Howard A., 41.

*redeem it. [Sterns shows us biblically that] the gospel, the
whole Gospel, means much more than the personal salvation
of individuals. It means a social revolution based on Christ's
work and the Spirit's power."[165]*

The popular evangelistic Gospel message of individual salvation and assurance of ultimate personal fulfillment in heaven suggests that "the Christian [salvation] narrative [focuses] less [on establishing] … the visible kingdom of God on earth and more [on] the [personal] journey from earth to heaven… Salvation becomes a movement from the material to the spiritual."[166] This suggests a separation of heaven from earth. Snyder points out that the separation of heaven and earth implies a hierarchical relationship of heaven over earth—the belief that the earth is temporary and heaven is eternal. From this evangelical reference frame, the value of the visible kingdom on earth depreciates. The salvation message becomes misdirected and solely focused on the individual, which fails to convey the whole meaning of salvation, that all creation is healed. The separation of heaven from earth misses the mark. "Jesus' incarnation and resurrection united heaven and earth."[167] Jesus is the "Lamb of God who takes away the sins of the world" (John 1:29). Sin has consequences for the world, humans, and the spiritual and material realms. Jesus' life-giving work reclaims what was lost at the fall: the sinless eternal union of the "dust of the earth" with the "breath of life" (Gen. 2:7).

The popular Christian point of view fails to comprehend biblical teachings about creation and God's purpose and plan for creation. The Gospel message was intended to be preached to "all creation" (Mark 16:15). With the separation of heaven from earth, the earth is viewed as a resource under human dominion, for human gain, not God's creation under human care, managed responsibly for human benefit. The separation of heaven from earth allows human economic and political preferences to distort and undermine God's plan.

Distorted Views of Heaven and Earth

According to Snyder, human ideas of nature and the earth have been distorted in the following ways:

---

[165] Sterns, Richard, *Hole in our Gospel*, (Nashville, Thomas Nelson, 2009), 17, 20.
[166] Snyder, Howard A., 9.
[167] Snyder, Howard A., 25.

1. <u>Some people romanticize nature as the arbiter of all outcomes</u>: Nature manifests deep beauty, mystery, and powerful forces that control and determine outcomes. Nature turns through the cycle of life and death—the power no human can overcome. Accepting this reality can lead to a fatalistic frame of reference—that is, we should enjoy life by indulging ourselves while we can until we reach the inevitable end. Alternatively, the beauty and power of nature reveal the character of the transcendent, intimate, indiscriminate God. Life and death and glimpses of life after death reveal God's purpose, meaning, and ultimate eternal outcome for humans and all creation, not fatalism.

2. <u>Some people see the earth as a source of raw materials to make and sell consumer goods</u>: Land may be divided up and owned as the property of private individuals, but the earth and all creation belong to God; the earth is not something to be exploited by humans for profit. According to Snyder, dominion and caretaking "means holding the earth in trust for all people, including the unborn generations."[168] God holds all humans accountable for their use or abuse of the earth. Humans are to be responsible caretakers of the created order.

3. <u>Some people worship nature</u>: They divinize the created order; creation becomes their God. According to Snyder, "New Age mysticism, various flavors of pantheism, even pantheistic forms of Christian theology [have blurred the distinction] between Creator and created. Nature, God, and ourselves become pretty much the same thing."[169] God created nature. Therefore, nature is not God. To worship nature is idolatry. Snyder says that nature worship can take various forms, including "worshiping ourselves, another person, our [material possessions], land, culture, [nation], and our [individual] rights..." If the primary focus of our worship is on these things, then "we are worshiping the creation, not the Creator."[170]

4. <u>Some people spiritualize the natural order</u>: They see the physical and material as sources of enjoyment that "lift us to higher spiritual truths." According to Snyder, under these conditions, the "material world has no value in itself; we do with it what we will, using and abusing it for our benefit, disregarding its integrity and well-being. This is dualism—cutting apart that

---

[168] Snyder, Howard A., 44.
[169] Snyder, Howard A., 44.
[170] Snyder, Howard A., 44.

100

which Scripture welds together."[171] The physical world is relegated to sustaining physical well-being and personal pleasure and teaching us the higher spiritual truths" about what to expect in heaven. This impoverished theology declares that "Christ rose physically to save us spiritually."[172]

The Creator's indiscriminate presence throughout the physical world can teach us spiritual lessons. The created order sustains our existence and has an intrinsic value, purpose, and destiny imparted by the Creator. "Physical things and life forms have their own right to exist because they come from [the Creator's word] and are overseen by [the Creator]."[173]

The Sevenfold Barrier[174]

Snyder explores the roots of the North American Christian Fundamentalist reference frame that characterizes creation care as "misguided or even morally subversive... a wicked political agenda that is anti-God, anti-American, and anti-free enterprise."[175] Christian fundamentalists view the Bible as authoritative, a source of God's truth that is literal and inerrant. Some conservative Protestant evangelicals embrace and promote the fundamentalist scriptural frame of reference. Fundamentalist groups, apart from Catholics, mainline Protestants, and Anabaptists, do not see "creation care as an essential part of the Good News." The modern-day fundamentalist worldview, explained by the four distortions described above, has progressed into a "sevenfold barrier" that distorts the biblical understanding of salvation and opposes creation care.

Snyder's sevenfold barrier helps to unpack how fundamentalists developed resistance to biblically-based creation care:

1. The theological inheritance of Greek philosophy: Hellenistic philosophy impacted church teaching in the second and third centuries. Snyder says, "Christian theology began to view the material world as separate from and strictly inferior to the spirit world."[176] Heaven is eternal and a place of perfection; earth is temporary and subject to decay. The Greek ideas of spiritual perfection and material imperfection became "deeply embedded in Western

---

[171] Snyder, Howard A., 44-45.
[172] Snyder, Howard A., 45.
[173] Snyder, Howard A., 45.
[174] Snyder, Howard A., 46.
[175] Snyder, Howard A., 46.
[176] Snyder, Howard A., 46-47.

theology."[177] The implication of this convoluted Greek-Christian theology is the distorted view that "the soul's journey from earth to heaven [is of greater importance] than the healing of all creation."[178] Snyder quotes N. T. Wright:[179]

> *Our minds are so conditioned... by Greek philosophy,*
> *whether or not we've ever read any of it, that we think of*
> *heaven as by definition nonmaterial and earth by*
> *definition as nonspiritual or nonheavenly. But that won't*
> *do. Part of the central achievement of the incarnation,*
> *which is then celebrated in the resurrection and the*
> *ascension, is that heaven and earth are now joined*
> *together with an unbreakable bond and that we, too, are,*
> *by rights, citizens of both.*

2. <u>The impact of the Enlightenment</u>: The Enlightenment period, known as the Age of Reason, lasted 200 years (1600-1800) and produced new challenges to religious teachings. Descartes, a mathematician, promoted the idea of mind-body dualism—he is known for his famous quote, "I think, therefore I am." Galileo founded modern astronomy, discovering that the planets orbited the sun and moons orbited the planets. Newton formulated the equations of motion and the law of gravity. Maxwell developed a set of equations that characterized light as an electromagnetic wave. Carnot started the science of thermodynamics, which led to the concept of a heat engine—burning coal to produce steam could run an engine that performed useful work. These discoveries yielded technological advancements that raised the human standard of living. Snyder stated that the material world, which humans already viewed as temporary, became a resource to be used and exploited for human gain, with few ethical limitations.[180] Since the earth was temporary, destroying life-giving ecosystems for human material benefit did not become a moral-justice concern.

3. <u>Embracing the economic ideology of capitalism</u>: Snyder says, "Capitalism became the [driver] of growth and prosperity... It brought material, economic, and often political benefits." Capitalism fueled industrialization,

---

[177] Snyder, Howard A., 47.
[178] Snyder, Howard A., 47.
[179] Wright, N.T., *Surprised by Hope: Rethinking Heaven, the Resurrection, and the Mission of the Church*, (New York, HarperOne, 2008), 251.
[180] Snyder, Howard A., 48.

leading to a high standard of living in Western countries. In the U.S., democracy and free enterprise provide for financial success.[181]

In truth, any economic system that humans run (feudalism, capitalism, socialism, communism, etc.) will be corrupted by human greed and exploitation. Jesus warns against storing up treasures on earth, yet some Christians promote prosperity as a sign of God's blessing for faith and pious behavior. Snyder states, "Surprising numbers of Christians have bought the central myth of capitalism: that the self-centered pursuit of profit inevitably works for the common good. This is very difficult to defend biblically."[182] Many Christians give the economy high priority and do not critique their role in the economic system. Uncritical acceptance fails to oppose individual immorality and systemic injustice. According to Snyder, some believe that "Since economic growth is good, the [unfettered use] of natural resources is morally necessary, not to be questioned... Many Christians oppose protecting the environment because they think this would place an unfair burden on business and stunt economic growth."[183] Responsible capitalism is a blessing. However, unregulated free markets can harm the earth's ecosystems, and an uncritiqued economic system can oppress and marginalize people.

Dualism informs us that material is inferior to the spiritual. Enlightenment teaches us to use the earth's resources for human gain. Capitalism promotes growth and progress to raise the standard of living. Earth's life-giving ecosystems become the victims of human "progress."

4. <u>American individualism</u>: Snyder states, "The 'rugged individualism' of North American culture undermines the sense of interdependence, shared responsibility for the common good, and earth stewardship."[184] The rugged individualist sees the beauty of nature and encounters it as a force to be reckoned with; some individualists may choose to live in harmony with nature, while others may choose to conquer it. American individual freedom has

---

[181] Those were not the only factors. Indeed, the unjust displacement of indigenous people from their native lands without compensation and the oppressive enslavement of African peoples against their free will provided the benefits of low-cost land and labor—essential contributors to the success of the 18th-century U.S. agricultural-based economy. Snyder states, "American Christians should, therefore, be cautious about claiming that God has uniquely 'blessed' America. It is a mixed and often morally muddy history." Snyder, Howard A., 50.

[182] Snyder, Howard A., 50.

[183] Snyder, Howard A., 50.

[184] Snyder, Howard A., 51.

fostered personal development and prudent risk-taking and delivered many social benefits. However, rather than striking a healthy balance of individualism with "social solidarity and mutual responsibility," the U.S. has cojoined individualism with "consumerism and materialism."[185] The corrupted individualistic value system normalizes greed and selfish ambition and fails to heed Jesus' teaching in Luke 21:15, "Watch out! Be on your guard against all kinds of greed; life does not consist of an abundance of possessions."

Balanced with interdependence, individualism fosters creativity and innovation, guards against all kinds of greed, and promotes sustainable and just creation care. Individualism distorted by greed fails to recognize the importance of the sustainable use of the Earth's resources. Some individualists who become prosperous view environmental regulations, which benefit life-giving ecosystems, as an affront to their freedom and an undue limitation on their prosperity.

5. Uncritiqued nationalism: Patriotism that inspires one to work for the good of their country and unites the nation's citizens is healthy and proper. Healthy nationalism recognizes a people group or nation's identity, independence, and sovereignty. Nationalism becomes unhealthy when it excludes others' interests, prioritizes national interests over people groups, and looks contemptuously at other countries. Any criticism of one's nation is viewed as unpatriotic. Uncritiqued nationalism becomes ideological and, in the extreme, cultish. Snyder says, "Uncritical nationalism leads to idolatry."[186] Christians avoid turning their love of country into idolatry when they see themselves first as citizens of the kingdom of God. Loyalty to their country comes second to their faithfulness to God's kingdom. I Peter 2:9 informs us, "But you are a chosen people, a royal priesthood, a holy nation, God's special possession, that you may declare the praises of him who called you out of darkness into his wonderful light."

Snyder emphasizes, "Genuine Christians see the whole earth from a global, not just national, perspective." This global perspective is a key aspect of the Christian faith, enlightening believers about the interconnectedness of all people and nations.[187] The priority of kingdom citizenship is not just a love of God but also the love of all people. Patriotic unity with one's fellow citizens

---

[185] Snyder, Howard A., 51.
[186] Snyder, Howard A., 52.
[187] Snyder, Howard A., 52.

must be viewed in the context of God's love not only for one's nation but all people of all nations.

6. Dualistic doctrine of salvation: Snyder says, "A biblical worldview demands that we pay attention to what the Bible teaches—not only about God's original creation but about the place of the created order in God's saving plan."[188] Some Christians focus on salvation through a personal relationship with Christ that produces a "personal new creation," assurance of salvation, and eternal life in the heavenly realm. However, this doctrine of salvation is incomplete; it focuses only on the spiritual part of God's creation and does not include God's complete plan and purpose for the world. This dualistic plan of salvation separates the physical world from the spiritual world. God's plan of salvation is all-encompassing, addressing both the spiritual and physical realms. John Piper affirms, "Christianity is not a Platonic religion that regards material things as mere shadows of reality."[189] God's comprehensive salvation plan gives believers reassurance and a sense of completeness. The physical resurrection of the body is as relevant and important as the soul's immortality. This emphasis on the bodily resurrection underscores God's value and significance in earthly existence.

The impoverished theology that Christ rose physically only to save us spiritually will not do. The incarnation, resurrection, and ascension united earth and heaven in body and spirit with an unbreakable bond. The Gospel message is to all creation, material and spiritual. Salvation is the healing of all creation.

7. Premillennial dispensation: Twentieth-century Christian fundamentalism has roots in a nineteenth-century evangelistic movement that popularized dispensational premillennialism. This eschatology, based partly on a "literal" interpretation of 1 Thessalonians 4 and Revelation 20:1-6, has gained wide acceptance. It teaches that Christ will return to earth to take Christians (living and dead) to heaven (rapture them) before the seven-year tribulation. Following the tribulation, Christ will return to establish the millennium, a 1,000-year reign of Christ on earth. During the time following the rapture, the Old Testament prophecies are fulfilled, and Jews come to faith. The final judgment follows this.

The idea of rapture in U.S. evangelical communities grew in influence in the second half of the twentieth century during the nuclear arms race and the

[188] Snyder, Howard A., 52.
[189] Piper, John, *Future Grace*, (Sisters, OR, Multnomah, 1995), 374.

threat of mutually assured destruction. For many Christians, the perceived nearness of the end times made the expectation of rapture feel like an imminent reality. According to Snyder, premillennialists believe that before Christ's reign during the millennium, "Society and world conditions will simply deteriorate. Since this is God's plan, there is little point in improving things now. Our exclusive focus should be on rescuing souls for the eternal future."[190] This theology calls for the evangelist to lead people to spiritual salvation, not to improve the lives of the unfortunate or to walk with the oppressed.

For the premillennialist, the earth's corruption and deterioration make its destruction inevitable. According to Balmer, the "Evangelical penchant for dispensationalism [results in a] lack of concern for the environment and the natural world... If Jesus is going to return soon to rescue the true believers and to unleash judgment on those left behind, why should we devote any attention whatsoever to the care of the Earth, which will soon be destroyed?"[191] This theological position rests on an interpretation of 2 Peter 3:10, "But the day of the Lord will come as a thief in the night; in the which the heavens shall pass away with a great noise, and the elements shall melt with fervent heat, the earth also and the works that are therein shall be burned up." (KJV). The NIV reads, "But the day of the Lord will come like a thief. The heavens will disappear with a roar; the elements will be destroyed by fire, and the earth and everything done in it will be laid bare [fully disclosed]." The premillennialist interprets "the heat and fire" as destruction. Alternatively, as Snyder puts it, "Fire is a symbol of God's power and holiness (Dt 4:24, 9:3; Heb 12:29), which can destroy if disregarded but is intended to cleanse all impurities (refine and purify—see Mal 3:2, Zech 13:9) so that people may experience and exhibit the pure love of God... Viewed in the full biblical context, the heat and fire of 2 Peter 3:10 signify refining, revealing, and cleansing, not destruction and annihilation."[192]

Premillennial dispensation advances the Greek philosophy of separating the spiritual from the material to an end-time rapture-tribulation event, where heaven is the eternal spiritual home of saved souls, and earth is temporary, subject to decay, and destined for destruction.

---

[190] Snyder, Howard A., 56.
[191] Balmer, Randall, *The Making of Evangelism: From Revelation to Politics and Beyond*, (Waco, Baylor University Press, 2010), 39.
[192] Snyder, Howard A., 59.

Chapter 7. A Theology of Creation Care – Section 2

Biblical Texts Opposing Dualism

Premillennial dispensationalists interpret Gen. 2:7 as God imparting an eternal soul to a physical body. According to Rydelnik and Vanlaningham, 2:7 refers to the process of God "imparting a *'soul'* [...which] allows humans to commune or 'relate' to God at a level that *transcends material creation...* Humanity and the creator uniquely share the *capacity for spiritual relationship...*"[193] This aligns with Greek dualism. The Greeks believed in immortality, but the human body was subject to decay and could not return to life once it entered the grave. Thus, they argued that the eternal soul was released from the mortal body upon death. This logic became the basis for dualism, eventually influencing the Christian doctrine of dualistic salvation.

A close reading of Gen 2:7, John 3:16-17, Is 55:12, Is 49:13, Mark 16:15, Col 1:23b, Lev 25:4, Rom 8:11, and Phil 3:21 indicates not a separation but the union of material and spiritual. Salvation heals and restores the material and spiritual. The kingdom has come near, and the work of salvation has begun. Resurrection is material and spiritual. Heaven and earth are not separate but in eternal union with Christ, the creator, reconciler, and restorer.

In Genesis 2:7, we read that God formed humankind from the dust of the earth and the breath of life. God forms a "living being" from the material and spiritual union. Arnold states, "The 'living being' is not some disembodied component of the human being, distinct from his physical existence; a 'soul' comprising one portion of a person's whole being. Rather, the 'living being' denotes the totality of the human."[194] God is intimate with the humans that God created. Humans are intimate with the earth, receiving physical bodies and sustenance. Humans are intimate with God, from whom they receive the breath of life.

John 3:16-17 has been used to proclaim a gospel supporting heaven's separation from earth. John 3:16 reads, "For God so loved the world that he gave his one and only Son, that whoever believes in him shall not perish but have eternal life." Looking closely at this text, the Greek root word for "world"

---

[193] Rydelnik, Michael, and Vanlaningham, Michael, *The Moody Bible Commentary: A One-Volume Commentary on the Whole Bible by the Faculty of Moody Bible Institute*, (Chicago, Moody Publishers, 2014), 40-41.

[194] Arnold, Bill, T., *Genesis*, (New York: Cambridge University Press, 2009), 57-58.

is *Kosmos*. This text does not read, "For God so loved humans..." If that were the case, the Greek root word would have been *Anthropos*, not *Kosmos*.

The Greek root word for "whoever" is *pas*, which means every part that makes up the whole, all things, everyone. The language in this verse includes all creation. The word *pas*, "all things," correlates with *Kosmos*, "all creation" that God made and called good.

In John 3:17, we read, "For God did not send his Son into the world to condemn the world, but to save the world through him." Looking closely at the text, we see that the Greek root word for "save" is *Sodezo*. *Sodezo* means to make well, heal, make whole, and save. Sin caused the brokenness in our relationship with God, each other, ourselves, and the earth. Human sin has caused harm to the order God created and called good. Salvation is all-inclusive. God created the world and humans and established the relationships between them. Salvation is the "healing," or "making whole," of all God created: the world, humans, and their relationships.

In Isaiah 55:12, we read, "The mountains and the hills before you shall burst into song, and all the trees of the field shall clap their hands." In Isaiah 49:13, we read, "Shout for joy, you heavens; rejoice, you earth; burst into song, you mountains! For the Lord comforts his people." In many places, the authors of the Old Testament scripture assign human-like qualities to inanimate objects. Trees clap their hands, and mountains sing for joy to the glory of God and the wonder of God's salvation. There is no dualism. The material and the spiritual are one. The Hebrew people saw salvation as the saving of all creation.

Mark 16:15 reminds us that Jesus said to his disciples, "Go into all the world [*Kosmos*] and preach the gospel to all creation [*ktisis*]." Colossians 1:23b reads, "This is the Gospel that you heard, and that has been proclaimed to every creature [*ktisis*] under heaven, and of which, I, Paul, have become a servant." The Greek root word for creation or creature, *ktisis*, means all creation, every creature. The good news message of salvation is to be preached to the whole of creation and every creature. Salvation comes by preaching the Good News to all creation.

God's intention to restore the earth does not begin on judgment day. In Leviticus 25:4, God declares, "But in the seventh year the land is to have a sabbath rest, a sabbath to the Lord. Do not sow your fields or prune your vineyards." Sabbath rest is for humans and the earth. Rest gives humans and the earth time to repair and regenerate. Therefore, strive to live into sabbath rest

108

and healing in the present. We connect to God, ourselves, others, and creation by actively participating in healing all creation in the here and now.

In Romans 8:11, Paul writes that God will give "life to your mortal bodies." In Philippians 3:21, Paul states that God "will transform our lowly bodies so that they will be like his glorious body." Christians in the early church believed God bodily raised Jesus from the dead, with a physical and glorified body, and that one day, they would be raised in the same manner. This was the fulfillment of becoming a new creation in Christ. Resurrection is material and spiritual, marking the eternal union of the dust of the earth and the breath of life.

Greek dualism, which separates the material from the spiritual, aligns with premillennial dispensationalism, which focuses on the soul's journey from earth to heaven, where the promise of salvation is personal fulfillment in heaven. On this basis, the faith community focuses on personal faith that produces assurance of salvation, not on the work for peace and justice to make real improvements in the here and now. The dualistic separation of heaven and earth is secular; scripture informs us of the material and spiritual union. Salvation is not the personal journey from earth to heaven but the healing of all creation. To misunderstand God's plan to heal all creation is to misunderstand the plan of salvation and the purpose of the kingdom coming near.

The Four-fold Alienation

Snyder informs us that all creation started as good, but "with sin came a moral disease, a fourfold alienation of man and woman from God, themselves, each other, and the earth."[195] The original sin (Genesis 3), a self-ascendant human desire to become "like God" (3:5), alienated the human relationship with God (guilt, avoidance—3:7a, 3:8, 3:10b, 3:11) and ultimately led to alienation from self (shame, fear—3:7b, 3:10a), each other (blame—3:12, 3:13), and earth (expulsion from the garden—3:23). The unanticipated outcomes of self-ascendancy did not make humans more like God but made them realize the inadequacy of self-reliance—"they sewed fig leaves together and made [inadequate] coverings for themselves" (3:7); "God made garments of skin for Adam and his wife and clothed them" (3:21). Humans realized a new vulnerability.

Self-ascendancy has a sad irony; our attempt to be like God led us away from God and towards self-reliance. The more self-reliant one seeks to become,

---

[195] Snyder, Howard A., 3.

the more profound the alienation from God, self, others, and the Earth. Humans' most profound need is a reconciliation of the four-fold alienation. Humans cannot solve this problem. God initiates reconciliation in God's way and time. God goes beyond reconciliation to complete restoration. Restoration includes the defeat of evil (3:15). "Salvation is the healing [reconciling and restoring] of all creation."[196]

According to Snyder, the Christian theology that incorporates the separation "between heaven and earth"[197] distorts human understanding of the Gospel and God's plan of salvation. The separation of heaven from earth theology must be reformed "to grasp the larger healing that salvation implies and promises."[198] Snyder lists the following symptoms as evidence of the separation of Heaven from Earth:[199]

- We see spirit and matter as opposing categories.
- We think salvation is about the soul only, not the body.
- We view death as the end of our earthly life.
- We view heaven as the place of our eternal spiritual life.
- We see the present world as evil, doomed to destruction.
- We overlook the biblical mandate for creation care.

Snyder says, "The Gospel is about healing the disease of sin."[200] Sin is at the root of the fourfold alienation, and salvation is God's plan to reconcile the alienation. Francis Schaeffer says,

"Christians who believe the Bible are not simply called to say 'one day' there will be healing, but that by God's grace substantially, upon the basis of the work of Christ, substantial healing can be a reality here and now."[201]

Christ's work aims to bring material and spiritual healing in the here and now. Snyder says, "Working to heal and reconcile the fourfold alienation based upon the provisions of the Gospel is indispensable in Christian mission."[202]

---

[196] Snyder, Howard A., 65.
[197] Snyder, Howard A., 3.
[198] Snyder, Howard A., 3-4.
[199] Snyder, Howard A., 4.
[200] Snyder, Howard A., 65.
[201] Schaeffer, Francis, A., *Pollution and Death of Man: The Christian View of Ecology*, (Wheaton, Tyndale House, 1970), 67.
[202] Snyder, Howard A., 68.

110

The Implication of the Four-fold Alienation

1. Alienation from God: Sin brings alienation between humans and God. When Adam and Eve sinned, they felt guilt and tried to avoid God by hiding themselves. Humans became isolated and self-reliant—a significant relational loss. The open, trusting, authentic, vulnerable, life-giving, intimate relationship with God was strained and broken. Guilt brought the need to hide and cover up, signifying the loss of security, peace, and joy. Loss of intimacy with God now brought the ominous possibility of eternal separation from God.

2. Alienation from self: Fear and shame, which followed alienation from God, brought alienation within ourselves. Self-alienation prevents us from knowing who God made us to be. This is a significant loss, as it prevents us from knowing our life's purpose and making meaning for our lives by living to fulfill that purpose. Snyder says, "Because of sin, people are not at home within themselves. The feeling of unmet inner needs brings uneasiness, disquiet, inner conflicts, and anxieties. We become double-minded; we outwardly appear one way but inwardly have alternate intentions—personal integrity is lost...[self-alienation] brings a whole range of maladies and symptoms that psychology and psychiatry deal with."[203] God provides resources for healing inner alienation through counselors, discipleship within the faith community, and the indwelling of the Holy Spirit. Inner healing, a lifelong journey, makes faith relevant, reveals one's purpose, and makes life's meaning evident.

3. Alienation from others: Alienation from God brings profound inner alienation; inner alienation brings alienation from others. Snyder says that alienation from God and self "multiplies alienation in the whole web of human relationships."[204] Humans replace authenticity and vulnerability, which bring joy and intimacy, with judgment and fault-finding, which bring defensiveness, pain, and loneliness. Rather than looking inward and self-evaluating, humans make excuses and launch accusations. Assignment of blame becomes the justification for physical, verbal, and psychological abuse against others. These behaviors deepen alienation from each other and can lead to violence and retribution. When trust is lost, suspicion of others follows, and mutually empowered, interdependent, life-giving relationships suffer. Individualism becomes a survival mechanism that seeks security through domination, power,

---

[203] Snyder, Howard A., 73.
[204] Snyder, Howard A., 71.

and control. The individualist carries inner fears and deep hurts, projects their fear onto others, and transfers the pain of relational woundedness to others. These behaviors harm intimacy and acceptance. Since we are created to be in relationships with each other, this is a significant loss.

4. Alienation from the earth: Outside the garden, humans must work the land to grow food in soil that produces thorns and thistles. Banishment creates a perceived threat of scarcity. This threat often leads humans to prioritize meeting material needs, sometimes at the expense of creation, furthering their alienation from the earth. As a result, humans value security, wealth, and material comfort over responsible use and care of the earth. Mismanagement of the Earth's resources harms vital ecosystems.

Additionally, the belief that physical bodies return to the material earth while disembodied souls ascend to spiritual heaven, even seeing that rapture will precede tribulation, intensifies this alienation. Paraphrasing Richard Stearns, the human focus on the afterlife reduces the importance of what God expects of us in this life.[205]

Competing Christian Belief Systems

Christians have a role in God's plan to reconcile the alienation from the earth. God placed humans in the Garden with a purpose: to till and keep. Those who view the earth merely as a resource for profit often see little value in healing humans' alienation from it, as this conflicts with their focus on prosperity. How can the entire Christian community work together to achieve reconciliation?

The original sin of self-ascendancy resulted in a realization of vulnerability. After the expulsion from the Garden, humans became self-reliant, which diverted them from loving God to loving material wealth, ultimately corrupting their hearts. Those with corrupted hearts "use the freedom and power they possess from self-ascendancy, often at the expense of others."[206]

Economic prosperity coupled with a corrupted human heart leads to storing up treasures on earth, which Jesus taught leads to more profound alienation (Matt 6:19-21, 24). Jesus' warnings about the moral consequences of enslavement to wealth are seldom preached in church. Some churches even promote becoming wealthy as a sign of moral goodness and blessing from God.

---

[205] Stearns, Richard, *The Hole in Our Gospel* (Nashville: Thomas Nelson, 2009).
[206] Snyder, Howard A., 49.

Economic progress has benefits, but the desire for wealth has moral downsides. Misuse of economic power can exploit poor people, silence their voices, and leave them trapped in the margins for generations.

The goal of personal wealth generation and ownership should be critiqued. Rather than being allowed to operate within the faith community, it should be examined for the exploitation of creation, the oppression of people with low incomes, and the indulgent lifestyle it may cause. "Uncritical acceptance of capitalism – giving it a moral pass – undermines the healing of creation... [When personal wealth generation and ownership are the goal], exploiting natural resources is morally necessary and not to be questioned."[207]

Many fundamentalist Christians promote an idealized morality of economic progress and oppose the liberal, "extremist" environmental agenda, which, in their mind, reduces profits, stunts growth, and creates a competitive disadvantage in the global marketplace. The Christian critique should call for a shift from personal wealth generation and ownership to Christlike wealth management and stewardship within the faith community.

1. Dualistic belief system: Platonic dualism, premillennial dispensa-tionalism, and the Western metanarrative of progress have merged with individualism over time to form a belief system that embraces capitalism's tenets of free markets, private ownership, and resource control. Material prosperity, at the expense of others and the earth, becomes acceptable within the normalized "excessively individualistic interpretation of the Christian message."[208] This belief system leads to an impoverished understanding of the purpose of being Christian, reduces Christian missiology to methods of saving individual souls, and alienates humans from the Earth. The dualistic belief system holds that a new creation will replace the existing earth; therefore, there is no need to work for sustainability and justice in the present age.

The dualistic belief system emphasizes restoring one's relationship with God but often neglects to address the deep wounds caused by a fourfold alienation. Material prosperity can lead to a self-gratifying earthly experience while anticipating a rapture into the perfect heavenly realm. In this perspective, the Earth is seen primarily as a resource for progress, which can be exploited for personal gain. Progress, for the minority wealthy, relegates the majority

---

[207] Snyder, Howard A., 50.

[208] Groome, Thomas H., *Christian Religious Education*. (San Francisco: Jossey-Bass Inc., 1999), 46.

poor to an inferior underclass.[209] Furthermore, the dualistic belief system deflects responsibility for the destruction of the Earth's vital ecosystems and overlooks the oppression of people with low incomes.

2. Reconciliation belief system: In the dualistic cultural-religious context, atonement for sin is often viewed as affecting only the redemption of the individual soul rather than the healing of all creation. In the reconciling belief system, salvation is "the healing of all creation."[210] Jesus' mission involves reconciling and restoring all creation by healing all alienated relationships between God, self, others, and the earth.

Christians have a role as servants of God in God's plan to reconcile the fourfold alienation. They critique individualized dualistic belief systems and shift to a reconciliation belief system.

Reconciliation of the alienation between God and humans comes not solely by claiming a personal relationship with Jesus but through a lifelong journey to reconcile the fourfold alienation: from God, self, others, and the earth. Self-aware, other-focused, community-oriented people engage in actions that promote this reconciliation.

Ecclesiology shifts from viewing the church as a collection of individual born-again believers who are assured of salvation, evangelize the lost, and worship a benefactor God, to understanding it as a community of faithful followers who work to establish the kingdom and testify to the gospel by living responsibly, loving mercy, acting justly, and caring for creation.

The atonement evolves from focusing on evading divine punishment for personal benefit to embodying an act of reconciling love, which calls Christians who receive God's mercy to love others as Jesus loved them.

The concept of the eschaton shifts from envisioning a raptured soul entering eternal spiritual security to a resurrected body and soul experiencing the eternal shalom of a reconciled heaven and earth.

The fourfold alienation of sin requires a fourfold reconciliation.[211] The Gospel reconciles humans with God, self, others, and the Earth. In this context, reconciliation replaces the dualistic belief system.

---

[209] Bauckham, Richard, *Bible and Mission: Christian Witness in a Postmodern World* (UK: Baker, 2005), 94-98.

[210] Snyder, Howard A., 65.

[211] Snyder, Howard A., 78.

According to Snyder, salvation in Christ begins with "substantial healing"[212] in the here and now. Followers of Jesus are called to advance the kingdom and do the work of reconciliation. We must work to reconcile the alienation of all broken relationships, which includes becoming caretakers of the earth until the time of the eternal shalom comes.

Theology Produces Action

Dula points out the necessity of arguing against "dominion theology and soul/body dualism that claimed salvation for humans at the exclusion of the rest of creation."[213] Wiebe's review article concurs with Dula; the "stewardship" model of creation care replaces the "domination by divine command" model. Each author advocates intentional movement toward a strategy to "overcome dualism" and "alienation between humans and non-human nature."[214] A holistic response to the climate crisis includes pragmatic scientific, economic, and political components to complement the theological and spiritual.

Timely and meaningful progress toward averting the climate catastrophe necessitates that all members of Christ's body take climate action. The church must find ways to engage and minister in the world without becoming like the world. Christians must begin by examining themselves. What prevents us from putting our faith into action? Christians must support faith-based programs and come alongside the work of secular organizations. All are invited to follow Christ in the work of reconciliation. How do you hear the call of Christ to participate in the healing of the four-fold reconciliation? How has the Holy Spirit led you to use your gifts in ways that bring God's complete healing to a broken world? Do you feel led to join the redeemed community and work for the salvation of all creation? Can you discern God's will and how to use your gifts to participate in God's plan in ways that please God? Later in Section V of the book, some examples of ways to take action will be discussed.

Christians must do more than pray. Work must be done at the personal and systemic levels. Christians witness kingdom values and priorities in the public square by facilitating and engaging in constructive conversations. This work involves advocating for mutually empowered stakeholders to build a collaborative constituency. The problem-solving mode calls for root-cause

---

[212] Snyder, Howard A., 66.
[213] Dula, Peter, 9.
[214] Wiebe, Joseph R., 357.

analysis, seeking the best knowledge, and implementing tangible, sustainable, and just improvements. The bias is toward problem-solving, not staking out positions and defending them in divisive debates. The problem-solving mode involves active listening and authenticity on behalf of all stakeholders, which leads to discernment on the way forward and implementation of practical solutions. The path forward includes the collaborative stakeholder group monitoring progress and identifying and making agreed-upon plan adjustments as needed. This process comprises achieving goals, celebrating accomplishments, and reaffirming each stakeholder's identity and role.

A case in point is the manufacture and use of lithium-ion (Li-ion) batteries. EVs get their energy from rechargeable Li-ion batteries. Electrification of the transportation sector has gained momentum in recent years. Consumer vehicles, small trucks, and light-duty service vehicles (such as U.S. Postal Service vehicles) are all seeing a rise in demand. Sales of consumer and light-duty plug-in EVs have increased from less than 2% to more than 10% market share in the past ten years.[215] The transportation sector in the U.S. produced 1.82 billion tons of GHG in 2019 (see Figure 7). Research has shown that the manufacture and use of EVs produce 60% lower GHG emissions than internal combustion engines (ICEs), which run on fossil fuels.

New lithium mines are being developed to meet the demand of the growing EV market, and existing mines are being expanded. Many known lithium deposits and other lithium minerals needed for batteries are located within or near U.S. tribal lands.[216] Social justice concerns for sacred areas and traditional ways of life must be considered. In addition, mining's impact on waterway pollution and biodiversity must be mitigated. The demand for lithium will increase as the U.S. transitions away from fossil fuels to meet the Paris Agreement's CO_2 emission targets. Environmental, mining, tribal, industrial, and government stakeholders must cooperate to build a collaborative constituency. On the one hand, Christians who act as caretakers of creation should advocate for a mutually empowered, interdependent team of stakeholders working together from start to finish of an EV battery project. On the other hand, the EV transition will take too long when battery manufacturing

[215] Argonne National Laboratory, "Light Duty Electric Drive Vehicles Monthly Sales Updates," *U.S. Department of Energy*, (updated September 2023), accessed October 17, 2023

[216] Block, Samuel, "Mining Energy-Transition Metals: National Aims, Local Conflicts," *ESG Research*, (June 3, 2012), accessed October 18, 2022.

programs have to jump through all the hoops, and fossil fuels are the assumed status quo. The sustainable and just transition to net-zero transportation must not make the perfect the enemy of the good. There isn't time for 10-year negotiation periods with every stakeholder. A balanced approach that is expeditious is required to meet COP26 goals. Karen Narwold, executive vice president at Albemarle Corporation, says, "We look very seriously at all aspects of environmental stewardship and community engagement. We're at the infancy of the lithium revolution, but we all know what's expected of us from a sustainability standpoint."[217]

Fossil fuel extraction, use, and burning cause more significant, irreversible harm to the environment, human health, and life-giving ecosystems than the manufacture and use of Li-ion batteries.[218] The Li-ion supply chain must be just and sustainable. Li-ion batteries that use cobalt have long cycle life, high energy density, and fast charging capability. The majority of the cobalt comes from the Democratic Republic of Congo, where unsafe working conditions and forced child labor have created an ethical and moral problem. In addition, people have been evicted from their homes, forests have been cut down, and water has been contaminated for the sake of mining. A recent report in the Mennonite Central Committee Peace and Justice Journal indicates that armed local groups fighting for control of mines and smuggling routes have displaced millions of people, many of whom require critical assistance.[219]

Agnes Callamard, Amnesty International's secretary general, has stated, "Amnesty International recognizes the vital function of rechargeable batteries in the energy transition from fossil fuels. But climate justice demands a just transition. Decarbonizing the global economy must not lead to further human rights violations."[220]

Changes are being made to fix the cobalt supply chain problems, but progress has been slow. Following the Amnesty International report,

---

[217] Milman, Oliver, "There's lithium in them thar hills – but fears grow over US 'white gold' boom," *The Guardian*, (October 18, 2022), accessed October 18, 2022.

[218] Hoffs, Charlie, "Challenges and Opportunities in Mining Materials for Energy Storage Lithium-ion Batteries," Union of Concerned Scientists, (updated December 22, 2022), accessed October 17, 2023.

[219] Alexander, Tammy, Editor, "U.S. Policy Towards DR Congo," *Peace and Justice Journal*, (Vol LVI, No. 3, Fall/Winter 2024), accessed October 6, 2024.

[220] Amnesty International, "Powering Change or Business as Usual," Amnesty International, (Updated September 12, 2023) accessed October 17, 2023.

automakers now select cobalt suppliers "that are audited for adherence to [safety] standards and that use cobalt only from mechanized industrial mines, where child labor is prohibited."[221] Research has demonstrated that iron can replace cobalt, lowering battery costs and increasing capacity. Battery recycling technology has shown that most cobalt, nickel, iron, and manganese can be economically recovered and reused, reducing the reliance on mining raw materials.

---

[221] Houreld, Katharine, "Clean Cars, Hidden Toll:," *The Washington Post* (Updated August 4, 2023) accessed October 17, 2023.

Chapter 8. Climate Justice

In this section, we present studies that quantify how people of low wealth and income and marginalized racial minorities are disproportionately affected by the energy industry and climate disasters—the consequences of climate change have both economic and health implications. The implication for Christians is to correct the policies that enable racially biased historical inequities and call for moral and just responses that make real improvements in the lives of those oppressed by systemic injustice. This section will conclude with a call to self-examination, repentance, and righting that which is wrong.

Those Who Contribute the Least Suffer the Most

Many Christians see climate change as a moral and justice issue. The climate crisis becomes a moral problem because it causes physical loss, migration, and conflict. It becomes a justice issue because those with low wealth and income, who have the lowest carbon footprint and do the least harm, suffer the most and have the fewest resources to deal with the effects. The immoral and unjust destruction of life-giving ecosystems opposes God's plan of reconciliation and salvation.

Samson et al. published a paper that quantitatively showed "that climate change will have widely varying effects on human well-being in different regions of the world."[222] The map shown below in Figure 23 compares, by country, the per capita emissions (upper map) versus the vulnerability to climate change (lower map). The map illustrates "Those who contribute the least greenhouse gases will be most impacted by climate change."

---

[222] Samson, J., Berteaux, D., and McGill, B.J., Humphries, M.M., "Geographic disparities and moral hazards in the predicted impacts of climate change on human populations," *Global Ecology and Biogeography: A Journal of Microbiology*, (Feb. 17, 2011), accessed 1/29/2024.

Figure 23: Per Capita CO2 Emissions versus Vulnerability to Climate Change[223]

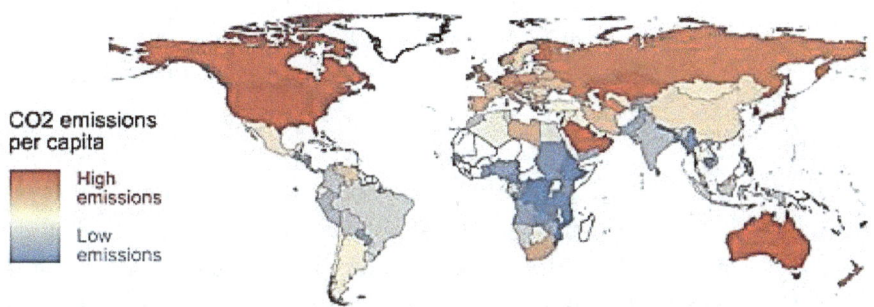

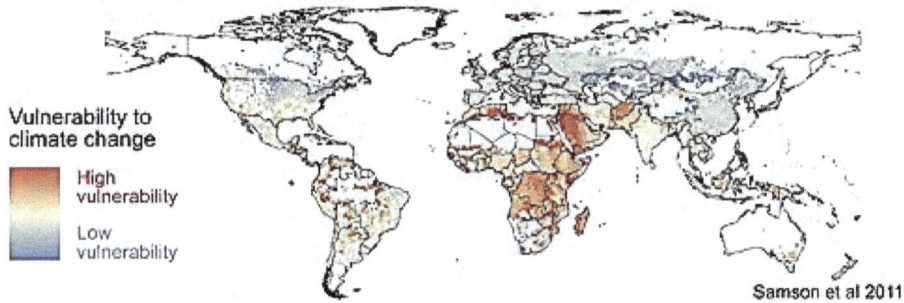

Samson et al 2011

Hayhoe says, "The poorest and most vulnerable will be most affected by climate change. That is why Christians are called to care about this issue... If we took our faith seriously, we would be at the front of the line demanding climate action rather than dragging our feet at the back."[224]

Examples of Disproportional Impact on the Marginalized

In the U.S., current energy production and fossil fuel consumption have already disadvantaged the most marginalized. Here, we will examine how racially biased housing policies have made low-income and minority populations disproportionately vulnerable to and unequally impacted by the harm caused by power plant emissions and hurricanes.

---

[223] Cook, John, "Those Who Contribute the Least Greenhouse Gases Will Be the Most Impacted by Climate Change," HuffPost, (Mar. 16, 2011 – updated Dec. 6, 2017), accessed 1/29/2024.

[224] Hayhoe, Katherine, *How to Save a Planet* Podcast.

Coal-fired power plants emit ozone, heavy metals, greenhouse gases contributing to climate change, and fine particle pollution (SO_2, NO_x) that causes acid rain and respiratory illnesses. Fine-particle pollutants ($PM_{2.5}$, particulate matter <2.5 μm) emitted from coal-fired power plants are breathed in through the lungs, enter the bloodstream, and are transported to vital organs. $PM_{2.5}$ pollution levels have been shown to cause premature death and are correlated to heart attacks, respiratory illnesses, hospitalizations, and lost workdays. $PM_{2.5}$ pollution contributes to climate change and is the 6th highest risk for early death globally, claiming 4.1 million lives annually (according to the Health Effects Institute and the Institute for Health Metrics and Evaluation's Global Burden of Disease Project). [225]

In 2005, the EPA issued rules to reduce $PM_{2.5}$ pollution from coal-fired power plants in the U.S. The energy industry brought lawsuits against the rule changes, which were struck down in 2008 for failing to conform to the Clean Air Act. In 2010, the Clean Air Task Force commissioned a study by Conrad Schneider and Jonathan Banks on the health effects and death rates attributed to $PM_{2.5}$ pollutants.[226] This study compared EPA power plant emission data with epidemiological studies to estimate the correlation between $PM_{2.5}$ emission levels and health outcomes. Prior large-scale investigations had documented a "direct link between power plant emissions and human health."[227]

The Clean Air Task Force study compared favorably with EPA studies and found that using scrubbers could provide a 50% reduction in $PM_{2.5}$ emission levels from coal-fired power plants. The estimated benefit from using scrubber technology was a reduction in the annual death rate from ~24,000 to ~13,000, a decrease in the number of heart attacks from ~38,000 to ~20,000 per year, and a reduction in hospitalizations from ~22,000 to ~10,000. These health improvements would deliver an estimated $100B in annual cost savings.[228] Many power plants oppose the installation of scrubber technologies because they are too costly to operate. Still, the health cost savings from the $PM_{2.5}$

[225] Bauer, Michael, "The State of Global Air/2018: A Special Report on Global Exposure to Air Pollution and its Disease Burden," *HEI and IHME* (2018): 1, (accessed May 25, 2023.

[226] Schneider, Conrad, and Banks, Jonathan, "The Toll from Coal: An Updated Assessment of Death and Disease from America's Dirtiest Energy Source," *Clean Air Task Force*, (September, 2010), accessed May 25, 2023.

[227] Schneider, Conrad, and Banks, Jonathan, 8.

[228] Schneider, Conrad, and Banks, Jonathan, 5.

emission reductions would help to offset the cost of running scrubbers on power plants.

"Deaths and illnesses [from PM$_{2.5}$ emissions] are major examples of coal's external costs, i.e., uncompensated harms inflicted on the public at large."[229] This burden is not distributed evenly across the population. Adverse impacts are especially severe for the elderly, children, and those with respiratory disease.[230]

In addition, low-income and minority populations are disproportionately impacted by the pollution from coal-fired power plants. Neighborhoods near power plants have lower home values. Power companies avoid locating power plants upwind of affluent communities.[231] The majority of people living within 3 miles of the "dirtiest" coal-fired power plants are people of color. "Minority groups… living downwind of [coal-fired] power plants are disproportionately exposed to the health risks and costs of fine-particle pollution."[232]

Next, we look at the impact of extreme weather. One tangible consequence of climate change is the forced displacement of people due to droughts, wildfires, rising sea levels, flooding, and hurricanes. Extreme weather events are becoming more frequent, and the associated costs are increasing. Climate scientists have shown that as atmospheric greenhouse gas levels continue to rise, so will the frequency and severity of natural disasters (floods, winter storms, tornadoes, hurricanes, droughts, and wildfires). The human and economic toll of extreme weather events has also increased, having a remarkable impact on people's livelihoods, causing large numbers of fatalities, destruction of homes, and damage to infrastructure.

Two case studies will be reported next: the impact of Harvey, a Category 4 hurricane, on the Houston area and the impact of Katrina, a Category 5 hurricane, on the New Orleans area.

In 2017, Hurricane Harvey caused severe flooding in Houston and surrounding areas, displacing more than 30,000 people. The flooding affected both high-income and low-income houses. According to the Brookings Institution report, the costs of Hurricane Harvey were "not evenly distributed."

---

[229] Schneider, Conrad, and Banks, Jonathan, 10.
[230] Schneider, Conrad, and Banks, Jonathan, 4.
[231] Global Energy Monitor, "Coal Plants Near Residential Areas," *GEM wiki* (April 30, 2021) accessed May 25, 2023.
[232] Schneider, Conrad, and Banks, Jonathan, 4.

Hurricane Harvey greatly impacted low-income and minority communities, and they struggled more to recover.[233]

The Brookings Institution report identified factors influencing economic outcome disparity post-Harvey. First, affordable, low-income housing was concentrated in areas of Houston prone to floods with substandard flood prevention measures. This factor increased the likelihood of flooding and the magnitude of the impact associated with storm-related flooding. The Brookings report went on to say, "Low-income and minority families are more likely to live closer to noxious industrial facilities and thus are more at risk to chemical spills and toxic leaks resulting from storm damage."[234] Wealthier neighborhoods in Houston were less likely to be severely impacted by storm flooding.

Second, low-income families had fewer financial resources to recover from home damage caused by flooding. Flood insurance costs have risen, and many cannot afford this coverage. Only "17% of homeowners"[235] living in the Houston area affected by Hurricane Harvey had flood insurance. Wealthier households' policies covered repair costs and the replacement of personal belongings. Those who did not have flood insurance had to rely upon "charity, government aid, and grants from FEMA, the Federal Emergency Management Agency."[236] The FEMA grants were challenging to get and capped well below the costs of home repair and replacement of personal belongings. Even if people could rebuild, their home value decreased since it was in a flood zone, resulting in a long-term financial impact: loss of wealth. Along with the loss of wealth came lower credit scores and greater difficulty securing low-interest-rate loans.

In 2005, Hurricane Katrina hit New Orleans; roughly 80 percent of the city was flooded, and "evacuation rates did not vary greatly across demographic groups."[237] Return rates were significantly affected by demographic groups. Census data from 2015 indicated that the Garden district (89.2% white, with an

---

[233] Krause, Eleanor, and Reeves, Richard V., "Hurricanes Hit the Poor Hardest," *Brookings Institution* (September 18, 2017), accessed May 25, 2023.

[234] Krause, Eleanor, and Reeves, Richard V.

[235] Long, Heather, "Where Harvey is hitting the hardest, 80% lack flood insurance," *Washington Post,* (August 29, 2017), accessed May 25, 2023.

[236] Long, Heather.

[237] Jayawardhan, Shweta, "Vulnerability and Climate Change Induces Human Displacement," *Consilience: The Journal of Sustainable Development*, 17, no. 1 (2017): 119.

average income of $125,000) had fully repopulated to the pre-Katrina levels. The Lower Ninth Ward (98.3% African American, with an average income of $38,000) had only 9.9% of its population return.[238]

Factors influencing post-Katrina return rates were evaluated for the Garden District versus the Lower Ninth Ward. New Orleans' post-Civil War, racially discriminatory housing practices "concentrated [African American] populations in ecologically vulnerable zones... African Americans were redlined into low-lying land."[239] In the 1900s, the US Army Corps of Engineers implemented flood protection programs based on home values. They provided better flood protection infrastructure for predominantly White districts with high home values. Higher-income white homeowners could afford flood insurance, greatly facilitating recovery and return, while lower-income Black residents could not afford flood insurance. The historical socioeconomic inequities of redlining, infrastructure, and flood insurance contributed significantly to long-term environmental displacement. "Marginalized groups had a greater exposure to the hurricane and less access to economic relief... after the hurricane."[240]

In this study of Hurricane Katrina, environmental displacement was identified as a socioeconomic problem. "Environmental displacement affects vulnerable populations disproportionately."[241] Marginalized people are more vulnerable to the long-term effects of displacement than wealthy people. Harvey worsened socioeconomic disparities.[242]

These studies show the impact of power plant emissions and severe weather events. The elderly, children, and people with respiratory illnesses are most vulnerable to $PM_{2.5}$ pollution from coal-fired power plants. Minority communities have the highest exposure to $PM_{2.5}$ pollution. Historical economic inequities of segregation have a disproportionate impact today on people of color. The systemic injustice of racially discriminatory housing policies, inadequate flood and chemical spill protection, and unaffordable flood insurance left low-income neighborhoods more likely to be flooded, with the fewest economic resources to recover from the flood damage and the most significant long-term negative impacts on wealth and livelihood.

---

[238] Jayawardhan, Shweta ,120.
[239] Jayawardhan, Shweta, 120.
[240] Jayawardhan, Shweta, 125.
[241] Jayawardhan, Shweta, 103.
[242] Jayawardhan, Shweta, 104.

124

The Call for a Christian Response

Christians are called to self-examine, repent, and do the kingdom work that rights wrongs. Christians are to walk with the oppressed, hear stories of the unfortunate, and work with them to establish justice and improve their lives. This requires correcting problematic frames of reference informed by political agendas and dismissing biased points of view that discredit the cry of the oppressed. Scripture calls for Christians to hold moral values consistent with biblical mandates and establish priorities leading to climate justice by caring for those most affected by climate change. By becoming merciful, pure-of-heart peacemakers, Christians live into their God-given life purpose to advance God's plan in a way that pleases God.

Self-examination calls us to ask: Do we act in accord with the prophet Isaiah when we see that the elderly, children, and those with respiratory illnesses are most harmed by coal-fired power plant pollution?

> *"Learn to do good; seek justice. Correct the oppressor. Take up the cause of the orphan; plead the case of the widow." (Is 1:17)*

We must repent of our privileged lifestyles and ask: Do we identify with Mother Mary's cry for the oppressed when the powerful and wealthy overturn life-saving air-quality rules and put economic interests above compassionate care for the unfortunate?

> *"He has performed mighty deeds with his arm; He has scattered those who are proud in their inmost thoughts. He has brought down the powerful from their thrones but has lifted up the humble. He has filled the hungry with good things but has sent the rich away empty." (Luke 1:51-53)*

We must do the kingdom work of preaching the gospel to those who have all that they need to live comfortably, discredit climate science, deny the effects of systemic injustice, ignore the oppression of the marginalized, and then claim to worship a God who is in control. Do they recognize that Jesus speaks to them when interpreting the prophet, Hosea?

> *"It is not the healthy who need a doctor but the sick. But go and learn what this means: I desire mercy, not sacrifice. For I have not come to call the righteous but sinners" (Matt 9:12-13).*

Jesus calls the "comfortable" to show mercy, walk alongside the unfortunate, and not demean them.

The energy industry, climate disasters, and socioeconomic disparities have health and wealth implications. The vulnerable, low-income, and marginalized populations are disproportionately affected.

Christians have a role and responsibility to act:

1. Christians call on leaders to correct the policies that enable inequities.
2. Christians debunk the false claims, disinformation, and zombie arguments.
3. Christians preach the Good News Gospel of salvation, the healing of all creation.
4. Christians correct the errors of dualism, which separates heaven from earth, focuses only on saving souls, and sees this world as doomed for destruction.
5. Christians do the work of the Kingdom of God (KOG) in the here and now.
6. Christians practice stewardship and creation care as a witness of the KOG here on earth.
7. Christians work as peacemakers, walking alongside the oppressed to support them in their work for justice.

Chapter 9. Ecospirituality in the Christian Faith Community

Here, we will explore the practices of self-ascendancy and the zero-sum game, and how they have deepened the four-fold alienation. We will use the Cain and Abel story as an alternative narrative to the American Experiment. We will discuss how ecospirituality informs the faith community's worship programs and offerings to invite personal transformation and meaningful social and systemic improvements.

The Cain and Abel Dilemma

We begin this discussion by reading Genesis 4, the Cain and Abel narrative. Using the framework of the Cain and Abel story facilitates consideration of how American narratives about the land, the Doctrine of Discovery, Manifest Destiny, and wealth continue to move us away from God's vision of the kingdom and how hearing and reflecting upon native wisdom might serve us better.

The bent towards self-ascendency, which began in the garden with Adam and Eve eating the forbidden fruit, continued, more disturbingly, between Cain and Abel after the expulsion. In Genesis 4:2b, we read, "Now Abel kept flocks, and Cain worked the soil." This difference in livelihoods anticipates the tension between the agriculturalists and the pastoralists in a post-Eden, pre-industrial world—they both need each other and are in conflict. Human flourishing depended upon the cooperation of the soil tillers and the livestock breeders, yet they competed for the same essential resource—the land.

Cain and Abel offer a sacrifice to God, an expression of gratitude in acknowledgment that the bounty of their work emanates from the provision of God's creation. They not only say who they are in their offering, but whose they are. Their physical substance comes from the earth, their work involves the earth's resources, and when done, they return to the earth. From birth to death, their very existence and purpose are the realization of the creator's initiative.

The Genesis 4 text does not explicitly reveal why God favored Abel's offering over Cain's. Instead, the text subtly contrasts the offerings. Cain brought "*some of the fruits* of the soil," and Abel brought "*fat portions* from some of the *first-born* of his flock." Abel willingly chose to bring his best. The New Testament (Hebrews 11:4) indicates that Abel's offering was an outward expression of his inner attitude—true faith in God and gratitude for God's provision. Abel's reverence for God appears to signify a turning from self-ascendency. On the other hand, Cain decided to give what was convenient and readily available. Cain's offering is not presented with humility, reverence, and gratitude, indicating Cain had not turned away from self-ascendancy. The text notes that Cain's offering did not please God.

In this narrative, there is no serpent, no one to tempt Cain. Cain chooses what to offer. Cain's reaction to God's disapproval is immediate and comes from within himself: "Cain was very angry, and his face was downcast" (4:5b). Cain was free to offer his best. He was also free to resist sin, "If you do what is right, will you not be accepted? But if you do not do what is right, sin is crouching at your door; it desires to have you, but you must rule over it," (4:7). Sin, for the first time, is mentioned in Genesis.

Cain had agency and knew right from wrong and good from evil. Cain was free to act morally or immorally, but his anger turned him from the path of reconciliation to deeper alienation. Cain failed to see that God provided for both, God loved both, and God's blessing was available to him and Abel. Perhaps Cain felt a competition for God's approval. Cain resented Abel receiving God's approval. Abel now had something that he did not have.

Cain did not learn from his brother's example of devotion to God. Cain did not examine himself, his motives, attitudes, or intentions; he did not resist evil. Instead, Cain chooses not to love Abel or show reverence to God. Cain's sin was premeditated. "Cain said to his brother Abel, 'Let's go out to the field.' While they were in the field, Cain attacked his brother Abel and killed him" (4:8). Cain willfully and knowingly committed murder. He remained unrepentant and knew there were consequences for rebelling against God and doing violence to others.

Arnold states,

> *"The prevalent characteristic of humans has become the propensity to harm each other. Life in the Garden of Eden was one of plenty for Adam and Eve, and there was no competition between them. Outside of Eden, humans reduced life to a zero-sum game. Whatever one individual [gains] is necessarily taken from someone else in this human reconfiguration of God's created order. Rather than producing new roles and relationships, humans [assume] that anything gained by one must be lost by another. Instead of a*

plentiful Eden in which humans are encouraged to be creative in their own right, humans have created a world in which they gain by taking from others."[243]

Following Cain's sin comes his denial of responsibility. "Then the Lord said to Cain, 'Where is your brother Abel?' 'I don't know,' he replied. 'Am I my brother's keeper?'" (4:9). Cain has added the sin of speaking falsely to his crime of murder. God makes clear that Cain alone holds responsibility for the deliberate act of violence and that Cain alone shall bear the consequences. In this case, the ground "will no longer yield its crops" (4:12). Following this consequence, Cain goes to the land of Nod, where he builds a city (4:17).

Abel demonstrates a right relationship with God. Cain's "anger" and "downcast" face (4:5) perhaps indicate Cain's jealousy of his brother's righteousness before God and shame for the inadequacy of his half-hearted offering to God. Cain's conduct reveals a heart that lacks devotion to God, harbors anger, and produces rebellion against God. It would seem that Cain's moral weakness allowed sin to rule his heart, leading to his failure to love his brother.

Self-ascendency led Cain into rebellion against God. He fell into the sins of pride and jealousy. The false notion of competition for God's approval led Cain to practice the zero-sum game. He wanted what Abel had and was shamed when God accepted Abel's offering, and rejected his. This shame produced anger. Cain's inner outrage was expressed in the outward act of premeditated murder. The sin of self-ascendancy and the practice of the zero-sum game deepened Cain's fourfold alienation from God, himself, others, and the earth.

The Cain and Abel narrative presents a dilemma: Rather than working at life-giving, reconciled relationships, humans justify personal gain at the expense of others.

What do Cain and Abel tell us about the dualistic, enlightened, capitalistic Western Europeans coming to indigenous people's lands in North America?

The indigenous peoples of North America used stone tools and lived tribal lives without the invention of the wheel and guns. Western European colonialists lived in an agriculturally based economy. They worked the soil and bred livestock. They built cities, had money, owned land, acquired wealth, made iron tools, and used technology. They established governments and nation-states.

---

[243] Arnold, Bill, T., *Genesis*, (New York: Cambridge University Press, 2009), 79.

Western European economies developed nation-states using a land-ownership structure. The flawed implementation of this political-economic system created wealth disparity and resulted in the unjust oppression of the marginalized. In the 17th, 18th, and 19th centuries, Western Europeans migrated to North America to escape financial and political system failures and seek relief from religious oppression.

The migration from Europe to North America presented opportunities and produced benefits for Western Europeans. The successes of the migration led some to see this as God's blessing and to testify that it was God's will. Western Europeans appropriated the Hebrew "Promised Land" narrative; they loved and obeyed God, who promised I "will bless you in the land you are entering to possess" (Deut. 30:16). They imagined the exodus from Europe to North America was a God-ordained entry and possession of the "new promised land" to establish the "new Jerusalem."

The migrants arrived in the New World and established their economic system based on their home country's agricultural and land ownership model. Similar to the biblical story of Cain and Abel, European immigrants and indigenous people competed for land. This created a human zero-sum game, resulting in winners and losers. Western Europeans required land and labor to ensure their financial success. The displacement of indigenous people from their ancestral lands without compensation and the enslavement of African people against their will allowed Western Europeans to obtain the land and labor they needed to cultivate crops and raise the animals vital for a prosperous economy. Their gain came at the expense of others.

Western Europeans established the Doctrine of Discovery to provide the moral and spiritual authority to proceed with their economic, political, and religious experiment. The Doctrine of Discovery declared that indigenous peoples were pagan and had no right to the land since they did not know Christ. Western Europeans used Biblical texts to justify the church and state's appropriation of land from indigenous peoples without compensation. The indigenous people were not under the authority of the church or the state government. Romans 13:1 states, "Let everyone be subject to the governing authorities, for there is no authority except that which God has established. The authorities that exist have been established by God." The Europeans saw themselves as the covenanted people of God with the mandate to take possession of the land. In Deuteronomy 1:21, we read, "See, the Lord your God has given you the land. Go up and take possession of it as the Lord, the God of your ancestors, told you. Do not be afraid; do not be discouraged." The indigenous peoples were viewed as the pagan infidels on the land God had given to the "chosen people" of Europe. The Manifest Destiny, which justified the Western expansion and further displacement of indigenous peoples, grew out of this text.

The enslavement of "heathen" Africans was understood by the application of the "Curse of Ham." In the Genesis narrative, Noah curses the descendants of Ham's son Canaan, "Cursed be Canaan! The lowest of slaves will he be to his brothers" (Genesis 9:25). This text was used to justify the enslavement of dark-skinned people. "Blacks were cast as Children of Ham, whom God had cursed to serve as slaves to Whites, the children of Shem and Japheth."[244] Slaveholders evangelized the enslaved as a justification for the loss of freedom in exchange for the salvation of their souls. The Virginia legislature in 1667 declared that the baptism of an enslaved person was "for the propagation of Christianity," but "did not exempt their bodies from bondage."[245] [246] Under the appropriation of body-soul dualism, baptism saved the enslaved person's soul, but their body remained enslaved. Therefore, the enslaved Africans became Christians while remaining held under the curse of Ham in bondage to the white owner.

The Declaration of Independence broke ties with the European monarchy, and the Constitution established democracy with guaranteed rights for its citizens, including freedom of religion. Taxation with representation, a central banking system, and fiscal and monetary policy were established to keep the economy running. The Doctrine of Discovery, the Curse of Ham, the Declaration of Independence, and the U.S. Constitution formed a "more perfect union" and "a nation under God." At that time, "citizenship" did not include indigenous or enslaved people groups or women. This meant that these people groups did not have the right to vote, own property, keep their earnings, or be given equal protection under the law. These doctrines and laws played a role in

---

[244] Mark, Joshua J., "Virginian Slave Laws and Colonial Development of Colonial American Slavery," *World History Encyclopedia*, (published April 27, 2021), accessed November 4, 2023.

[245] Educational Broadcasting Corporation, "Slavery and the Making of America," *Public Broadcasting Service*, (published 2004) accessed November 4, 2023.

[246] Encyclopedia Virginia, "Laws of Virginia," *Library of Virginia*, (published 2020), accessed November 4, 2023.

forming the ideas of American Exceptionalism, which fostered the policies that enabled the system of white European supremacy to function. The strongest nation on earth today has emerged from this experiment. Is this what God intended? Could this experiment have achieved more by following a different path?

Perhaps one may consider that God sent the Western Europeans to North America not to displace Indigenous people, enslave African people, or establish an economic system that enabled the human zero-sum game, but to turn from their self-ascendant ways and establish a reconciled relationship with the Indigenous people of the earth. If Western Europeans chose to leave their promised land narrative and dualistic thinking, which justified displacement and enslavement, they could have developed life-giving relationships with the land and the indigenous peoples. Instead, they moved in, played the zero-sum game, and created winners and losers. The Western Europeans missed the opportunity to witness Christ and discover new, joy-filled relationships with Indigenous peoples. Each people group could have benefited by learning from the other—establishing God's shalom.

Unlike Western Europeans, the North American indigenous people were not informed by philosophical dualism. They did not govern and own land the way Europeans did. The indigenous peoples retained their physical and spiritual communion with the land, which they saw as the source of their daily bread, and they remained in communion with the Creator, whom they worshiped as the maker of the land. Perhaps the actual Biblical call of God to the Western European migrants was:

1. Practice a responsible relationship with the land.
2. Enter a life-giving relationship with the indigenous people.
3. Turn from self-ascendency towards devotion to God.
4. Turn from the zero-sum game.
5. Grow the kingdom of God through just actions in the way of Christ.

The "new Jerusalem" was to be a "city on a hill." God's people were to be a light to the world. In the city on a hill, God intended the people to be witnesses of God's kingdom. "And this gospel of the kingdom will be preached in the whole world as a testimony to all nations" (Matthew 24:14). Western Europeans created a new world like the old one, which they gained by taking from others. The human zero-sum game distorted "God's calling" to be a "city on a hill" and corrupted the truth of the Gospel message. Sadly, instead of becoming a faithful witness of the kingdom, seeking to heal the fourfold alienation, they deepened alienation with the sin of self-ascendency at the expense of the vulnerable.

The American experiment has produced blessings for many. Freedom of religion, democracy, and economic opportunity have relieved many people from oppression, persecution, and poverty. Did this experiment have to come by way of the zero-sum game? Did the religiously and politically oppressed Europeans become America's religious and political oppressors? Does the success of the American experiment justify acting with impunity towards indigenous and African peoples? What role does American Exceptionalism have in continuing the fourfold alienation? What must be done to reconcile and restore life-giving relationships with God, self, others, and the Earth?

What is the sign that the kingdom has come near? How do we practice what Jesus calls us to do: "Love your neighbor as yourself" (Mark 12:31), "Do unto others as you would have them do unto you" (Luke 6:31), and "Love one another. As I have loved you, so you must love one another" (John 13:34). As citizens of the kingdom, we no longer seek self-ascendency and no longer practice the zero-sum game. Instead, we work to heal the fourfold alienation and establish life-giving relationships with God, self, others, and the Earth.

Cain tried to cover for his sin by denying that he was his brother's keeper, implying that everyone must fend for themselves. Can Western Europeans deny moral and ethical responsibility for the indigenous people's loss of land and the African people's enslavement? Biblical texts were used to justify control and ownership of the land and people. Biased beliefs that identified the non-Christian as "heathen," "inferior," "savage," and "dangerous" were used to justify policies that enabled systemic injustice and oppression. The "heathens" were "God-destined" to be displaced and enslaved. The threat that "dangerous heathens" presented was dealt with violently. The ideologies justified killing innocents to create and advance an exploitative economic system.

Is America the new Jerusalem, the city on a hill that is a light to the nations? Does our prosperity and success show that God has blessed us for our faithfulness? Or, does the American experiment's failure to acknowledge the moral consequence of violent acts that shed innocent blood put the entire system at risk of a downfall leading to a "great crash" (Matthew 7:27)? Is climate change a consequence of the American experiment? Will the environmental destruction that outpaces the ecosystem's capacity for

133

restoration ultimately result in an economic collapse?[247] How do we transition from practicing the zero-sum game and self-ascendency to healing the fourfold alienation?

Worship and Offerings

Ecospirituality, drawn from spiritual reflection, turns us away from self-ascendancy and the zero-sum game toward healing the four-fold alienation. Ecospirituality is expressed in worship and produces offerings. Our worship of God must be done with devotion to God, who knows and loves us. Our offerings to God must be an expression of gratitude, acknowledging that the bounty of our work emanates from the provision of God's creation. Our worship and offerings say who we are and whose we are. Our worship and offerings express reverence, a desire to turn from self-ascendancy, and alignment with God's plan for healing all creation. Our existence, purpose, and meaning in life are the realization of the Creator's initiative.

Spiritual reflection enables an ecospirituality that encourages the reconciliation of the fourfold alienation by caring for each other while caring for the earth. Ecospirituality in the faith community recognizes the union of the earth and heaven, the natural with the supernatural. We are the dust of the earth and God's breath of life. Worship enlivens our material and spiritual beings and communion with God. Offerings to God transform what we believe into what we do by hearing and practicing Jesus' words. Worship and offerings emanate from community discernment to guide us in healing the fourfold alienation. We discover Spirit-guided, joy-filled, life-giving relationships with God, self, each other, and the earth. We learn the beauty of God's creation more deeply and worship God with greater reverence, awe, and joy.

Worship and offerings move us beyond individualism and dualistic salvation and foster relationships with God, self, others, and the earth. By connecting the physical and spiritual in community, faith, and action, we live into our life's purpose and discover its meaning.

Ecospirituality, drawn from spiritual reflection, awakens our curiosity, interprets mysteries, and reveals the connection between the breath of life and the dust of the earth. We become a faith community whose members avoid the zero-sum game and shun self-ascendency. We remove "my way is right"

---

[247] Meadows, Donella, et al., xxv.

legalisms, do not seek to control, and prefer working together through mutual empowerment according to each one's gifts over hierarchical control.

Ecospirituality-based worship and offerings guide us out of the exceptionalism and supremacy of fundamentalist Christianity that justifies claim, conquer, and control. Examination of self-destructive ways leads to lament, grief, processing of shame and guilt, healing, renewed convictions, and the birth of life-giving relationships with God, self, others, and the earth. The harm caused by nationalism, exceptionalism, and supremacy can be healed. God and country are replaced by God and faith community. We participate in God's salvation plan, preparing us to enter the eternal shalom.

Ecospirituality-based worship and offerings energize everyone's role in learning and growing people. Mind, heart, spirit, and body are fully engaged in healing the fourfold alienation. Ecospirituality involves participating in the responsible healing of creation by practicing creation care and taking climate action at the personal, community, and systemic levels.

We journey along the way that Jesus leads by hearing and practicing his words: "Love the Lord your God with all your heart and with all your soul and with all your mind and with all your strength" (Mark 12:30), "Love your neighbor as yourself" (Mark 12:31), "Do unto others as you would have them do unto you" (Luke 6:31), and "Love one another. As I have loved you, so you must love one another" (John 13:34). These teachings become the moral and ethical guides to building life-giving relationships with God, self, others and the earth. They lead to the reconciliation of the fourfold alienation.

Ecospirituality-based worship and offerings bring the future just and righteous kingdom into the here and now. We acknowledge the immoral and unjust destruction of life-giving ecosystems that oppose God's plan of reconciliation and salvation. We examine ways we are complicit in the climate crisis, lament, and repent. We correct problematic frames of reference informed by economic worries and political agendas and dismiss biased points of view that discredit climate science.

We become Christlike, live sustainably, live justly, and bring good news to the marginalized. We become merciful, pure-of-heart peacemakers, willing to suffer for the cause of justice. We deliver real improvements to the oppressed, hope to those who have little to hope for, empathy and care to those who suffer loss, and restoration to those who have been displaced.

Ecospirituality-based worship ends the desire for self-ascendence and turns us away from dualistic Christianity. Ecospirituality-based offerings to aid the unfortunate provide an effective alternative to self-gratifying consumerism. Ecospirituality worship and offerings extend ecological care into morality and justice concerns:

1. We care for the people at the margins.
2. We hear the voice of the oppressed.
3. We declare the way of our Creator, who calls us to salvation—the healing of all creation.
4. We show mercy to the unfortunate and become kingdom-building coworkers with our Creator.
5. We do not demean the unfortunate but walk alongside them as merciful, pure-in-heart peacemakers.

By entering through this narrow gate and following the Creator's spiritual wisdom, we can carry out the work God has prepared for us. This allows us to make real improvements that help heal the fourfold alienation and bear the fruit of righteousness and justice. Through our perseverance and resilience, we embrace our God-given life purpose and contribute to God's plan for healing all creation in a way that is pleasing to Him.

Section IV

The Church Has a Message of Hope

138

Chapter 10. Looking Inward

This section explores resources and methods to develop spiritual reflection and informed decision-making skills. We examine internal personal hurdles and outward social and systemic challenges. What follows are some tools for embarking on that journey.

Futility and Decay Do Not Have to Become Fatalism, Despair, and Denial

For many, climate change causes a fatalistic attitude, despair, or denial. These attitudes can lead to a loss of hope, cynicism, or an attitude that it is already too late. The antidote to fatalism, despair, and denial is to know that everyone can make a meaningful difference for the better. Here, we will see how action produces hope, and hope inspires action.

In Paul's message to the Romans, chapter 8:18-22, Paul provides a theological understanding of the scientific fact that creation is "subjected to futility" and "enslaved to decay." "Futility" means no process is 100% efficient. Everything that humans do produces waste. "Decay" means things will deteriorate over time. If we make and use something, we must maintain it, or it will wear out, fall apart, and become useless. There is no way to eliminate futility and decay; they govern all we do. This is the lament of our existence. This fundamental fact of life on earth can lead to fatalism, despair, and denial.

Some hold a fatalistic attitude that assumes acceptance of undesirable but inevitable things. Fatalism can lead us to become indifferent to the problems of the world around us. A fatalistic attitude about climate could lead to enjoying the earth for what it provides while it still lasts, and not worrying about it beyond that.

Fatalism "quenches" the inspiration to act and "destroys" our sense of the God-given purpose for our lives. This leads us to think that what we do is meaningless and that we passively accept the inevitable outcomes of climate change. However, Christ promises an abundant life that offers hope and gives purpose. Purposeful living calls us to climate action because it creates hope that it is not too late to reverse the direction of climate change.

For some, the inescapable nature of futility and decay leads us from indifference to despair. We go from feeling like anything we do is meaningless to feeling hopeless. Hopelessness can produce inaction. The present difficulties are a fact of Christian life in the age between Christ's resurrection and Christ's return. Paul reminds us that "the sufferings of this present time are not worth comparing with the glory about to be revealed in us" (8:18). The struggles of this age are not an end but a prelude to glory. We do not let the difficulties of this present life drive us into despair.

The promise of salvation is full, life-giving communion with our creator. We live in the hope that God, who subjected creation to frustration, will one day set it free from "its enslavement to decay" (8:21). The magnitude of climate change may seem overwhelming and desperate. Rather than giving up in despair, we engage in the struggles of caring for the earth; we focus on what we know and can do. By taking meaningful climate action, we do our part to make real improvements.

Rather than being indifferent or hopeless, some may deny that humans are complicit in the climate crisis or that climate change is real. Denial saves us from worry and "useless" speculation about a climate catastrophe. Some believe this fallen, broken, sinful earth is disposable; it will one day be replaced by a better one. The best we can do is trust that God will provide and wait for the day Christ returns; then, we will be raptured into a new heaven and earth. Unfortunately, denying the consequences of climate change will not make them go away.

Paul is teaching us that because the issues of this age persist, we must not fall into denial. Paul reminds us that God "foreknew" (8:29) the ultimate destiny for our lives; God's plan of salvation "predestines us to be conformed" to the image of Christ (8:29). This teaching calls us out of denial. We live into our anticipated future by doing the works God has prepared for us today (Eph 2:10). This means that instead of denying the problems of climate change, we identify root causes and work to solve them.

Fatalism, despair, and denial affect our response to God's blessing to have responsible dominion over all living things (Genesis 1:28) and the command to be tillers and keepers of the earth (Genesis 2:15). God's provision carries humans through the futility and decay that govern all creation. Rather than falling into the trap of fatalism, denial, and despair, humans become responsible caretakers working to avert the coming climate catastrophe; everything humans do according to God's plan has meaning and purpose.

The Call of Moses[248]

Have you ever wondered, "Can I make a difference?" Or have you received a "call from God?" The idea of a divine call with a purpose that can significantly impact one's life is compelling. Yet, many of us find ourselves in moments of doubt, feeling stuck, lost, and uncertain. We question our gifts, importance, and worth, and wonder if God is truly at work in our lives. Does what we do or say genuinely matter? Can we, in our unique way, make a meaningful difference?

Perhaps the story of the burning bush (Exodus 3) may help. In Exodus 3:2b, we read, "Moses saw that though the bush was on fire, it did not burn up." This strange, unexplained sight stimulated Moses' curiosity. God got the attention of Moses. In this narrative, we notice God's call. In 3:4, we read, "When the Lord saw that he had gone over to look, God called to him from within the bush, 'Moses, Moses!'" At this point, you may say, "Stop right there." Moses received a clear call from God. God audibly spoke Moses' name, and Moses answered, "Here I am." God declared Moses to be on "holy ground" (3:5). How can this text be helpful? After all, you have never seen a burning bush.

The most striking aspect of this narrative is who calls first—not God, but God's enslaved people in Egypt. The enslaved people in Egypt cried out in their suffering. Moses was too far away to hear them, but God heard them. God's call to Moses directly responds to the people's cries. "I have heard them crying out... I am concerned about their suffering. I have come to rescue them..." (3:7-8). In this, we see God's deep empathy for the oppression and injustice they endure. God's call to Moses is a call to service, a call to be the instrument through which God will bring about his plan. This part of the narrative is crucial, leading to a pivotal moment—Moses' response.

From the burning bush, God's call to Moses revealed God's plan for Moses' life. You would think that Moses knew precisely what God wanted him to do. After God spoke from the burning bush, did Moses jump right up and say, "Yes, Lord, send me?" Not even close. Moses appears to resist God's call on his life. It seems Moses was content with his life as a shepherd. He was far away from the troubles he had in Egypt. He had a wife and a son and got along with his father-in-law. Moses made excuses and objected to God's call:

1. "Who am I, that I should go?" (3:11).
2. Who should I say you are? (3:13).

---

[248] Ott, Dan, "The Call of Moses," Eastern Mennonite Seminary Chapel, 8/29/2023, adapted from sermon notes.

    3. "What if they do not believe me?" (4:1).

    4. "I am not an eloquent speaker… I am slow of speech…" (4:10).

    5. "Please send someone else" (4:13).

The original sin was self-ascendency—attempting to be like God by eating the forbidden fruit. After eating, humans realized that they were naked. They had not become like God; they became afraid and ashamed. They hid, denied, and blamed. Outside the garden, self-ascendency produced a sense of scarcity and inadequacy. In Moses' list of objections, we hear his claims of not having enough authority (scarcity) and not being capable of the work (inadequacy).

God does not directly respond to Moses' feelings of not having enough and not being good enough. God does not affirm Moses as the best person for the job. God does not encourage Moses that he is capable enough. God does not declare, "You can do whatever you put your mind to." Instead, God promises constant accompaniment, "I am with you" (3:12).

Perhaps God does not refute or affirm Moses' objections because Moses is correct in his self-assessment. In reality, we are like Moses. On our own, we do not have enough, and we are not enough. We cannot end oppression and injustice. We cannot even stop our greed, bickering, and jealousy. Moreover, we cannot stop climate change. No person has the resources or the skills to do that. Moses did not have enough, nor was he good enough. We do not have enough and are not good enough.

Moses was promised that God would be with him and provide what he needed to do the job God had prepared for him.

It appears this promise is not enough. Moses seeks further assurance. How will the people know who you are and that you are with me? God gives a confounding response, "I am who I am" (3:14). This statement hardly seems to satisfy Moses. What could "I am who I am" possibly mean? How does this prove to people that God is with me or sent me? Moses does not comprehend the meaning of God's answer. On the one hand, God assures Moses of intimacy by promising, "I will be with you." On the other hand, the character of God and the nature of God's presence are shrouded in mystery with the declaration, "I am who I am." God seems to say: I am close to you and beyond your understanding. I know you, and someday, the mystery of who I am will be revealed. Now go and serve.

Moses continues to object, thinking he does not have enough and is not good enough, and wonders if God realizes who God is asking to do this job.

After hearing God's response to each of his objections, Moses appears to continue in doubt, as indicated by his final request: "Please send someone else" (4:13).

Rather than give his doubt to God, could it be that Moses wanted to hear God's words of affirmation, support, and encouragement? "I drew you to the burning bush because I have confidence in you. You will have everything you need to do the job. I guarantee your success."

Instead, Moses hears God's plan, in which God declares, I am the one who hears the cry of the oppressed people. I am the one who will save and deliver my people. I spoke to you to set you free from your doubt, and I will work through you to set my people free from their brutal enslavement.

Moses felt lacking and inadequate, and wanted to stay in his comfort zone. He was not ready to push beyond the boundaries of his limitations. He did not want to hear God's call. He needed proof of God's reliability and dependability.

Moses thought the call from God was about himself. Instead, it was about the oppressed people enduring brutal enslavement in Egypt and God's plan to work through Moses to save, deliver, and heal them. Moses thought he had to do it all, but was capable of nothing. God was the one who accomplished everything. Moses was the conveyer, the worker of God's plan, while God was the one who saved, delivered, and healed.

Climate change is a very large and complex problem. No one person has enough or is capable enough to solve the climate crisis. The magnitude of the potential catastrophe of ecosystem destruction is overwhelming. So, back to the first question: Can I make a difference? This is the wrong question. Moses could not see how he could make a difference. God was not asking Moses if he thought he could make a difference. God was asking Moses if he would go. The people's call for salvation became God's call to Moses.

God hears the call of those who suffer climate injustice, and God knows the suffering that will come upon future generations from the climate catastrophe. God's plan of salvation includes the healing of all creation. Has the climate crisis gotten your attention? Do you see how God can work through you to do the job of creation care, ecojustice, and ecospirituality? Can you give your doubt to God? Do you believe God is with you? Do you think the accomplishments of your work are God's, not yours? The question is not whether you have enough or are capable enough, but whether you are willing and ready to leave your comfort zone. Your "enough" is your obedience to the call, not your skill set.

We know we do not have enough and are not good enough, but we know we are not alone. "I am with you" gives us what we need to do the job. We realize that though we do not comprehend the mystery of "I am who I am," we know who is sending us. In doing the job God has for us, we discover who God made us to be and fulfill God's purpose for our lives. In that, we discover the meaning of our lives, and the mystery of "I am who I am" is revealed to us. Ultimately, we realize we had enough and were good enough because God made it so.

Chapter 11. Seeing Self and Others Clearly

Successful climate action requires connecting in community with others. Divisive debates produce more heat than light and rarely lead to meaningful change. This section provides suggestions for successful group conversations with a facilitator and a community of participants. In general, the conversation facilitator must come well-prepared. Ensure the participants know something about the facilitator and take time for introductions. Then, begin with an overview of the topic at hand. Give an outline of the discussion material and start with an opening question that people can easily engage in. The facilitator should be friendly and use a welcoming voice inflection that conveys enthusiasm and positive energy, stimulating curiosity. As the conversation proceeds, the facilitator must practice empathetic listening; everyone should be valued and treated with respect. The facilitator may want to share personal stories and demonstrate appropriate vulnerability, which can foster a sense of connection and openness. The conversation must be invitational. All present in the community are learners, including the facilitator.

Transitioning from Divisive Debates
to Constructive Conversations

The role of the facilitator is crucial to establishing healthy conversations with all participants. Healthy conversations should produce critical reflection and lead to informed action. People enter conversations at different starting points regarding their readiness to transition from feelings of alienation to reconciliation. Constructive conversation prompts participants to examine biases and assumptions. It reveals the effort required to distinguish between correlations and actual cause-and-effect relationships. Along the way, participants may encounter disorienting dilemmas that can lead to emotional processing. Increased knowledge and awareness foster understanding and promote reconciling behaviors. This work is challenging but essential. The primary goal of this work is to facilitate personal and relational growth, resulting in life-giving relationships, critical reflection, and informed action.

1. <u>Begin with respect</u>: The setting for constructive conversation must be a safe environment where mutual trust may be earned and respect given. Personal attacks, mockery, quick judgments, know-it-all advice-giving, or criticism of others are boundaries that must not be crossed.

Conversations may produce positive or negative energy. They stimulate the rational and emotional parts of our minds, our heart's desires, and the affections of our spirits. Conversation energy must be harnessed to reveal new knowledge, create a deeper understanding, and produce new awareness that informs new behavior for all participants.

Harnessing conversation energy in a way that builds trust and earns respect requires the conversation to take place in a safe environment.

2. What is a safe space? A safe space is one where participants feel free to share their points of view, where alternate opinions are accepted, and where people are respected and valued. Encouraging participant input through shared dialogue fosters a healthy communication process. A facilitator guides the direction of the dialogue to ensure a safe space that fosters learning through honest and respectful group dialogue.

Acknowledging all perspectives and avoiding criticizing or disputing participant responses is essential to creating a safe space. When participants struggle with concepts, follow-up comments and questions can encourage deeper interpretation and possible application. It is crucial to value participant responses, allowing for a lively dialogue that fosters community among participants.

This approach can help participants become invested in the conversation and feel comfortable asking questions and sharing personal struggles. By fostering a safe environment that stimulates productive dialogue, participants can feel heard and supported in their journeys within the conversation group.

3. What happens when safe space boundaries are crossed? When a participant engages in a divisive debate tactic that violates safe space (attacking others, becoming defensive, judgmental, critical, etc.), the conversation facilitator must respond to the disruptive comment confidently. The facilitator should approach the situation with curiosity without putting the participant on the spot. The facilitator should respectfully confirm they heard the comment correctly by repeating the participant's words. Rather than becoming defensive or justifying a personal point of view, the facilitator must become aware of their emotional response. They should quickly ask themselves what feelings I am experiencing and where they are coming from. Then, the facilitator should consider what the participant is trying to communicate about themselves. The facilitator can ask for additional clarification or information to better understand the participant's point of view. The facilitator may also provide

context or ask for input from other participants. The facilitator may also make connecting statements to prior group dialogue or ask for input from other participants. Even if a participant makes a misguided comment or violates the safe space, it should still be treated as a learning opportunity, not a divisive debate opportunity.

This approach creates an inclusive, safe environment for all participants to feel valued and to contribute to productive dialogue. Facilitators invite all participants to contribute and commit to learning together, fostering a sense of fellowship among participants on their unique journeys.

4. Tell personal stories: The safe environment creates opportunities to form connections where conversation participants get to know each other by telling their stories. Our stories unite us more than divide us. We listen to each other's joy and loss, pain and peace, with open minds and empathetic hearts. We feel less alone when we are better known and accepted as we are.

By using the power of the story, the conversation facilitator may become a role model for authenticity and vulnerability. Vulnerability and authenticity are foundational to establishing trusting connections that facilitate personal growth for all participants. Constructive conversation enriches our lives by helping us listen carefully to each other. We realize we all have something to say, learn, and need each other.

Appropriate personal storytelling models vulnerability and authenticity. Others listen and validate what they hear. This begins to establish a safe space. By being authentic, participants become known to each other. People form connections through sharing, hearing each other's real-life experiences, and understanding feelings. Respectful storytelling and attentive listening allow interpersonal connections to create and honor a safe space for constructive conversation.

In this environment, the Spirit activates caring connections, and the power of God's grace and truth becomes manifested. The Spirit's positive energy transforms the safe space into a sacred space. Caring connections build trust. We become known to each other and accept each other "as we are." We do not have to hide. When we are fully known, we are fully understood. These connections become the core of a community that practices constructive conversation.

5. Have patience: When a participant chooses to remain hidden and refrain from being open about themselves, they may be experiencing fear, doubt,

frustration, or hurt. This hiddenness serves as both a protective measure and a form of imprisonment. By staying hidden, they can never be fully known or understood by others, as they only allow others to see the parts of themselves that they are willing to reveal. Ironically, this choice to remain concealed makes them even more vulnerable to the negative emotions of fear, doubt, frustration, and hurt that confine them.

The walls the participant builds to hide their feelings ultimately create a barrier of isolation from God, others, and themselves. This self-imposed isolation prevents meaningful connections. By distancing themselves, they block the closeness they need with others. The walls that conceal their emotions also mask their fears and frustrations, which they may unintentionally project onto others. Ultimately, they may lose the one thing they desire most: interpersonal connection.

Please practice patience and avoid letting their projected negative energy divert the conversation into a divisive debate. Approach the situation with respectful curiosity. Ask questions about the person's experiences and feelings related to those experiences. Give participants who continue to display judgmental and critical behavior the opportunity to move from justification to exploration. When they feel safe and ready, they will open up, let down the barriers, and invite closeness. This process can take time, so allow for genuine connections to develop and take shape.

6. Transforming a relationship starts with making changes from within oneself: Making healthy changes takes time and effort. Betty Pries states, "The transformation of one's interior condition [is prerequisite to the transformation of] an interpersonal relationship."[249]

Let go of the need to have things go how you want them to. Tell yourself, "All is well. I am safe."[250] When you encounter a difficult person or conversation, welcome the feelings this situation produces. Do not respond with judgment or defensiveness. Do not ignore or avoid the situation. Do not repress your feelings. Welcome them. Name them, process them, and put them into context. See the situation from a perspective of wholeness and lovingness and through the eyes of Christ. Do not empower your feelings by fighting or

---

[249] Pries, Betty, *The Space Between Us: Conversations about Transforming Conflict*, (Harrisonburg, Herald Press, 2021), 16.

[250] Clymer, Donald and Clymer-Landis, Sharon, *The Spacious Heart: Room for Spiritual Awakening*, (Harrisonburg, Herald Press, 2014),134.

running from them; know where they come from. Rather than feeding energy to the feelings, heal them at the source.

When we take the time to listen to ourselves, we become more aware of ourselves. Knowing ourselves allows us to experience healthy love of self and see ourselves more clearly. When we see ourselves clearly, we see others and hear their stories. By knowing their stories, we can understand and have compassion for them. We can then choose to love them as Christ loves us. In doing so, life-giving relationships are born. There is nothing better than to be known and loved.

How to Have Constructive Conversations about Climate

Climate conversations quickly become divisive debates. How do we overcome the political, religious, economic, and social polarizations that create alienation, distrust, and disrespect? People put up defenses, mount attacks, and demonize those who oppose their position. Climate conversations can produce fears that lead to fatalism, denial, and despair. Protective walls are built to shut out unwanted inputs.

Divisive climate debates use misinformation to make claims, heightening the division between participants. Fake experts promote pseudoscience to slander actual experts. False assumptions lead to invalid beliefs. Bias occurs when the only data selected supports a claim. The rest is ignored. False assumptions, invalid beliefs, and bias inform problematic frames of reference. Digging in and defending problematic frames of reference increases ideological polarization. Listening does not occur; failing to hear one another leads to divisiveness, deepening division.

One strategy is to shift from trying to talk over each other to listening to each other. The purpose of listening is not to conjure up a rebuttal but to discover how what is being said makes you feel. Then, take the time to process those feelings. Where do these feelings originate? What do these feelings indicate about unmet needs? Personal feelings (not rebuttals, judgments, correctives, or criticisms) may be shared in a safe environment with authenticity and vulnerability. There is no judgment of feelings. Feelings are not right or wrong.

Hearing and coming to know the other's feelings may confirm (without evaluation) that they listened to the person correctly. Having done so, they may validate that they understand how the other person feels (without judgment).

This communicates empathy and makes the person feel like they have been heard. People want to know they have been heard before they want to listen to what you say. At this point, participants may begin to make connections and move away from alienation. The possibility for constructive conversation begins to emerge.

Unprocessed feelings become a barrier to constructive conversation. In a divisive debate fed by unprocessed feelings, we think we are being rational, making defensible claims supported by valid facts. We are unaware that we are emotional and fail to see how those emotions inform our thinking. In addition to not seeing ourselves clearly, we do not see or hear the other person. Unprocessed emotions energize divisiveness.

In a divisive debate, we become fixated on proving our point, declaring ourselves right, and labeling others wrong. Emotionally, we want "what we argue for" to be true because to think otherwise would feel devastating. The fear of losing an argument drives us to pursue victory at any cost; this prevents constructive conversation and making a connection. Ironically, the relentless desire "to win at any cost" hinders us from achieving what we deeply need: a life-giving relationship with each other.

After listening, validating, processing feelings, and letting go of the need to be right, we may also take the time to listen to God's voice. Hearing God and hearing the voice within us is crucial to being able to listen to each other. Our communion with God and self is vital to our communion with each other. We deepen our communion with each other by hearing God's voice. Our relationships with God, self, and others become essential to our caretaking of God's life-giving creation.

Bring People Together and Set Goals

When bringing people together, set some goals for a constructive conversation about climate change.

Goal 1: Commit to overcoming political, religious, economic, and social differences by respectfully speaking one's point of view, carefully listening to others' points of view, personally processing one's feelings, and respectfully validating each other's feelings. Show that you care about each other as people more than you want to change their minds about something.

<u>Goal 2</u>: Commit to overcoming fatalism, denial, and despair with action based on the best knowledge. Accept the challenge and learn by trial and error. Inspire each other to act, knowing that "mistakes" are not "failures" but "learning opportunities."

<u>Goal 3</u>: Commit to seeking the best knowledge from reputable sources of information and using all relevant facts and data. As much as possible, examine multiple sources, rely on recognized experts from current peer-reviewed research, and evaluate both sides of the argument to eliminate bias and correct false assumptions and invalid correlations.

<u>Goal 4</u>: Commit to honoring a safe space by being vulnerable and authentic, developing trust, and always being respectful.

<u>Goal 5</u>: Commit to personal emotional processing and validating what you heard the other person say *before* speaking.

<u>Goal 6</u>: Commit to speaking with respectful curiosity. Seek to understand the other person, not to refute or defeat their argument.

<u>Goal 7</u>: Commit to growing and learning together, using critical reflection and taking informed action. Know that what we do matters.

Following these goals can prevent one from defending one's point of view while dismissing others, allow one to have constructive conversations, and build meaningful relationships.

Understand Why Climate Science is Rejected

Hayhoe says people's rejection of climate science "is rarely about science itself... [People's opinions about climate change depend upon one's] values, ideologies, worldviews, and political orientation." [251] In addition to holding fast to the security of one's identity, people naturally resist change. Identity and resistance to change drive "tribalism." Tribalism reinforces our differences and increases our suspicion of those outside our tribe. The more our identity is challenged, the more threatened we feel, "the tighter we draw the circle to distinguish between them and us." [252] Divisions become emotionally energized and are not based upon rational thinking. Divisive debates produce accusations and defensiveness. People find safety and acceptance in their tribe. Like-minded views are cheered, and those with opposing views are discredited and

---

[251] Hayhoe, Katharine, *Saving Us:* 6.
[252] Hayhoe, Katharine, *Saving Us:* 6.

152

demonized. Feeling secure and safe becomes more important than growing personally and relationally.

Connect Who You Are to Why You Care

We discover what we have in common through constructive conversation in a safe space. This moves us from distrust and disrespect to collaborative problem-solving.

Hayhoe raises the importance of connecting *"who we are to why we care"* through storytelling.[253] Parents, conservationists, business owners, patriotic people, devoted religious people, nature lovers, outdoors people, gardeners, and farmers each act in ways consistent with their identity and values. Sharing personal lived experiences helps people learn about their interests and concerns. Your interests and concerns reveal something about who you are and why you care about the earth. Who you are informs your beliefs about climate change, and your values form the basis for why you care. What you do reflects what you care about.

Climate change does not distinguish between democrats and republicans, liberals and conservatives, seculars and the religious. Discovering what we have in common draws us out of our partisan tribes and short-circuits divisive debates. We find shared concerns about climate change and shared values for climate action.

Where do you live? What brings you joy? Whom do you love? What do you do for work? What social group do you come from? What is your faith tradition? What activities do you like doing? What are you passionate about? Whoever you are, your identity, interests, and concerns qualify you to discuss your climate change perspective. Hayhoe says,

> *"We care about the food we eat and how much it costs;*
> *how clean or dirty the air we breathe is; the economy and*
> *national security; hunger, disease, and poverty across the*
> *planet; the future of civilization as we know it. We've*
> *woven a million reasons why we already care about*
> *climate change into the very fabric of our society. We just*
> *haven't fully realized it yet."[254]*

---

[253] Hayhoe, Katharine, *Saving Us:* 19.
[254] Hayhoe, Katharine, *Saving Us:* 32-33.

We all care about climate change because it affects our lives, and the way of life and livelihoods of future generations. We should not try to change others into something they are not; instead, we should focus on discovering our shared values. We all want the Earth to be a safe place for humans to flourish and be hospitable to everyone for generations. To make this discovery, we need to create an authentic, vulnerable, safe space for constructive conversation where we can engage with our minds, hearts, and spirits.

Each time we work to transform a divisive debate into a constructive conversation, we transform conflict into the joy of a life-giving relationship. Healing relational alienation is the plan of salvation and "a deeply meaningful way to follow Jesus."[255]

---

[255] Pries, Betty, 12.

154

Section V

A Call to Action

Chapter 12. Use Your Heart, Mind, Spirit, and Body

The following section of this book briefly describes how we may act as caretakers of creation. These sections call for a spiritual reflection that identifies improvement areas and takes informed action that uses "best knowledge" to make real improvements at the personal, communal, and systemic levels. The call to action is holistic and involves our minds, hearts, spirits, and bodies.

Welcome Divine Grace

Theologian Randy Maddox states, "The change God effects through faith is rooted not only in intellectual assent to a set of doctrinal propositions but, even more, includes the affective transformation of the whole person through lived experiences of divine grace and love."[256]

Steve Thomas, the leader of *JoinTrees to Save the Earth*, states, "As St. Teresa of Avila said, 'Christ has no body now but yours. No hands, no feet on earth but yours.' We must pray, and we must act. And now. To cool the planet, plant a tree."[257]

Timothy R. Eberhart, holder of the United Methodist Church Chair in Ecological Theology and Practice, states,

> *"Faithful witness to God's mission for life must necessarily include an ongoing, rigorous accounting of humanity's personal, institutional, and structural sins against God, self, others, and the goodness of the created world. Walter Brueggemann teaches that pain can be a necessary precursor to hope and that mourning, when embraced, can lead to praise. That lament, individual and communal, is the spiritual source of genuine prophetic action. 'It is precisely those who know death most painfully who can speak hope most vigorously.'*

---

[256] Maddox, Randy, "John Wesley's Precedent for Theological Engagement with the Natural Sciences," *Wesleyan Theological Journal* 44, no. 1 (Spring 2009): 23–54.

[257] Thomas, Steve, "My Prayer: God Save the Earth," *Anabaptist World*, (August 26, 2022), accessed September 2022.

158

The logic Brueggemann affirms here is the scriptural logic of hope in God's power for life over death. Hope, throughout the scriptures, is not rosy optimism. The witness of hope in the scriptures begins with a tough realism: on the cross, in the tomb with a dead body, in exile by the waters of Babylon, in Egypt in bondage; it begins ex nihilo amidst the dark chaos covering the face of the deep. The purpose of facing the terrible possibility of the "ecological collapse of the earth" is not to demoralize spirits or demotivate action. Quite the opposite, the purpose is to awaken the deep energies of new life that can only come by turning directly into the truth—about ourselves, about the world—and making ourselves open to the creative movements of God's regenerating Spirit."[258]

How can we utilize personal experiences, self-examination, and critical reflection to transform new awareness into understanding and further into informed action? How can we steer clear of naive optimism while avoiding the depths of despair? This section offers insights into becoming informed and empowered by God's Spirit. By God's grace, we recognize our role as caretakers of creation, fostering a resilient, problem-solving mindset and realistic actions and new behaviors.

---

[258] Eberhart, Timothy R., "New Birth toward a New Earth Regenerative Mission for Planetary Renewal," *Missio Dei and the United States: toward a Faithful United Methodist Witness*, Kathryn M. Armistead editor, (USA: The United Methodist Board of Higher Education and Ministry, 2018), 69-85.

Pray

Use Prayers from *Voices Together*[259]
> **Leader:**
> "Now you are the body of Christ, and each one of you is a part of it"
> (1 Corinthians 12:27).

> **All:** *[Voices Together 1035]*
> Christ has no body on earth but yours;
> Yours are the only hands with which Christ can do his work.
> Yours are the only feet with which Christ can go about the world.
> [Yours are the only ears that can hear the cries of the oppressed.]
> Yours are the only eyes through which Christ's compassion can shine forth
> upon a troubled world.
> Christ has no body on earth now but yours.

> **Leader:**
> "Jesus said, 'You give them something to eat.' The disciples answered, 'We
> have only five loaves of bread and two fish.'" (Luke 9:13).

> **All:** *[Voices Together 1036]*
> Lay one brick at a time,
> Take one step at a time;
> We can be responsible
> Only for one action
> Of the present moment.
> But we know that God will take them
> And multiply them,
> As Jesus multiplied
> The loaves and fishes.

---

[259] Kauffman, Bradley, General Editor, *Voices Together*, (Harrisonburg, MennoMedia: 2020).

Leader:

"A new command I give you: Love one another. As I have loved you, so you must love one another. By this everyone will know you are my disciples, if you love one another." (John 13:34-35).

All: *[Voices Together 1037]*

Love cannot remain by itself,
It has no meaning.
Love has to be put into action,
and that action into service.
Whatever form we are,
Whatever our abilities,
Whatever our resources,
It is not how much we do,
But how much love we put in the doing;
A lifelong sharing of love with others.

Sing: "Here I am, Lord"   *[Voices Together 545]*

From Pope Francis:

Leader reads aloud: Prayer from Pope Francis
A Prayer for Our Earth [260]

Leader:
All-powerful God, you are present in the whole universe
and in the smallest of your creatures.
You embrace with your tenderness all that exists.
Pour out upon us the power of your love,
that we may protect life and beauty.

---

[260]Catholic Climate Covenant, "Pope Francis: A Prayer for Our Earth," Franciscan Monastery, (Jul. 30, 2024), accessed Aug. 8, 2024.

All:
Fill us with peace, that we may live
as brothers and sisters, harming no one.

Leader:
O God of the poor,
help us to rescue the abandoned and
forgotten of this earth,
so precious in your eyes.
Bring healing to our lives,
that we may protect the world and not prey on it,
that we may sow beauty, not pollution, and
destruction.

All:
Touch the hearts
of those who look only for gain
at the expense of the poor and the earth.

Leader:
Teach us to discover the worth of each thing,
to be filled with awe and contemplation,
to recognize that we are profoundly united
with every creature
as we journey towards your infinite light.

All:
We thank you for being with us each day.
Encourage us, we pray, in our struggle
for justice, love, and peace.

From Shane Claiborne:

Blessing of the Land or a Garden[261]

God of the Universe,
you made the heavens and the earth,
so we do not call our home merely "planet Earth."
We call it your creation, a divine mystery,
a gift from your most blessed hand.
The world itself is your miracle.
Bread and vegetables from the earth are thus also from heaven.
Please help us see your presence in our daily bread.

Upon this garden
may your stars rain down their blessed dust.

May you send rain and sunshine upon our garden and us.
Grant us the humility to touch the humus,
that we might become more human,
that we might mend our rift from your creation,
that we might then know the sacredness of the gift of life,
that we might truly experience life from your hand.
For you planted humanity in a garden
and began our resurrection in a garden.
Our blessed memory and hope lie in a garden.

Thanks be to God,
who made the world teeming with a variety,
of things on the earth, and under the earth.
Thanks be to God
for the many kinds of plants, trees, and fruits
that we celebrate.
For the centipedes, ants, and worms,
for mice, marmots, and bats,
and for the cucumbers, tomatoes, and peppers.

---

[261] Claiborne, Shane, et al., *Common Prayer: A Liturgy for Ordinary Radicals*, (Grand Rapids: Zondervan, 2010), 561-2.

We rejoice
that we find ourselves eclipsed by the magnitude
of generosity and mystery.
Thanks be to God.

Sam Hamilton-Poore, Presbyterian Church USA pastor and professor of Christian Spirituality at San Francisco Theological Seminary, states,

> *"Christian love in action on behalf of the earth's [life-giving ecosystems] is essential... and that action needs to include the work of human prayer. How we live is informed and shaped by how we pray, and how we pray is informed and shaped by how we live... Prayer helps us remain rooted in hope [and action]."* [262]

Prayer inspires us to see creation as a profound blessing and a sacred trust. Write your own prayer asking for God's guidance in using the Earth's resources sustainably and justly.

Worship

Earlier in this book, we discussed how God's plan of salvation is to heal all creation spiritually and materially. In the Old and New Testaments, creation expresses joy and mourns. In Isaiah 55:12, we read, "The mountains and the hills before you shall burst into song, and all the trees of the field shall clap their hands." In Isaiah 33:9, we read, "The land mourns and languishes." Romans 8:22 reads, "We know that the whole of creation has been groaning as in the pains of childbirth right up to the present time." In Isaiah 49:13, we read, "Shout for joy, you heavens; rejoice, you earth; burst into song, you mountains! For the Lord comforts his people and will have compassion on his afflicted ones."

Nature has a Divine purpose; it worships God in its enigmatic ways, and its beauty is a testament to God's glory. When humans live in alignment with the Divine purpose, we, too, worship and glorify God. However, when humans act irresponsibly, they disrupt God's created order and inflict harm on life-

---

[262] Hamilton-Poore, Sam, *Earth Gospel: A Guide to Prayer for God's Creation*, (Nashville: Upper Room Books, 2008), 10-11.

sustaining ecosystems. In this way, humans are responsible for the damage inflicted on creation. Scriptures affirm that nature and humans worship, suffer, and anticipate the marvel of God's salvation. The material and the spiritual are intricately intertwined. Humans are in a symbiotic relationship with nature. The Hebrew people perceived salvation as the redemption of all creation.

Outdoor worship helps us reconnect our spirits with our bodies and with nature. Janzen states, "We are invited to pay attention and enter worship with an expectation that God is present and active in the world around us and that creation itself has something to offer us in expanding our understanding of God."[263] "Ever since the creation of the world, God's eternal power and divine nature, invisible though they are, have been seen and understood through the things God has made. So, they are without excuse" (Romans 1:20).

Wendy Janzen, pastor of Burning Bush Forest Church, has written a series of worship services for outdoor worship. These outdoor services, which occur once per month, year-round, seek to bring people together to "nurture their souls and pay attention to God."[264] When people gather for forest worship, they are reminded that "they worship with creation, not just in creation."[265] These worship services are not mere replicas of our indoor practices. Worshipping in nature heightens our consciousness and redirects our focus to God's creative, awe-inspiring, and enigmatic aspects. Jansen affirms, "We acknowledge that in nature, nothing exists in isolation, but everything is intricately connected with other elements of the [life-sustaining] ecosystem. We acknowledge the significance of our connection with larger systems beyond ourselves."[266]

Take Personal Actions

Climate change is caused by humans burning fossil fuels. Corrective action begins with identifying the root cause and making necessary changes to resolve the issue. We tend to resist change because it requires work and takes time. Often, we are unsure of the outcome. In the case of climate change, we know the root cause – burning fossil fuels, we see the impact – global warming, and we know the corrective action – dramatically reducing fossil fuel consumption.

---

[263] Janzen, Wendy
[264] Jansen, Wendy, "About Us," Burning Bush Forest Church, (Updated December 2023), accessed December 12, 2023.
[265] Janzen, Wendy
[266] Janzen, Wendy

So, *we all* must identify ways to reduce our fossil fuel consumption. None of us can do everything on the list given below, but all of us can do some of the things that are listed.

Take a close look at your lifestyle. Understand how you consume fossil fuels. Ask yourself, "What is a sustainable and just standard of living? Make changes to reduce your consumption.

Take corrective actions by making informed decisions; use the best knowledge, seek the wisdom of God's Spirit, and hear the counsel of others in your community. Embody! Put what you know and profess to believe into action and understand that what you do makes a difference.

Work for systemic changes. Environmentally sustainable and just economic development requires the coordination of consumers, businesses, faith groups, and governments to achieve a shared vision of the Earth's environmental future. Each constituency has a crucial and significant role as a stakeholder in averting climate catastrophe; responsible consumerism, technological innovation, spiritual wisdom, targeted government regulations, and economic incentives contribute in essential ways. Christians must act as responsible consumers, wealth managers, and stewards of resources.

Work for Systemic Changes

Voters must support and elect politicians promoting a sustainable transition to a renewable, clean energy economy. We must not put climate deniers or those who oppose climate action into public office. Who we vote for and what we do make a difference. We must ask: Who are the leaders that give climate action the exposure and priority it demands? What climate action programs do we support? Which climate actions are we personally willing to take?

There is still time to take action that will have a significant beneficial impact.

1. Act now. Care for the earth before the "unquenchable fire" is ignited, and it becomes too late to save our way of life and souls.

2. Find your way forward and go. Move out of fatalism, denial, and despair and move into action. Dula stated, "Hope is not the prerequisite for action; it is the product of action."[267]

---

[267] Dula, Peter, Eastern Mennonite Seminary Chapel presentation, *Eco Theology*, April 19, 2022.

3. <u>Use your agency</u>. Identify specific things that you can do now. Seek the best knowledge and wisdom. Communicate with others. Develop a plan. Implement it. Learn as you go. Do not become locked in paralysis of analysis, skepticism, or pessimism. "Mistakes" are not failures; they are learning experiences. Find climate-friendly, greenhouse-gas-reducing actions that you can take. Evaluate these actions, try to update, improve, and share them with others. Change what you do and how you live. Let your climate-friendly lifestyle become a new normal and inspire others.

4. <u>Be an evangelist for creation care</u>. Collaborate with those who are alarmed and concerned. Reach out to the cautious and the disengaged – help them overcome their hesitancy and move forward on climate action and creation care.

Chapter 13. Use Best Knowledge and Take Informed Action

Practice Strategic Thinking
1. At the personal level, stop thinking what you do is too little or too late and start doing. Act, and know that the product of action is hope. Use the power of your mind and body.
2. At the interpersonal level, be a wise listener and knowledgeable speaker when relating to those who are climate deniers. Use the power of your ears and voice to connect with others.
3. At the systemic level: Know who you are voting for. Do not elect those who oppose climate action. Use the power of your vote. Advocate for meaningful policy changes. Hold the coal, oil, and gas industry, automakers, and power companies accountable for their GHG emissions. Make your economic choices based on their climate action performance. Use the power of your pocketbook.

Identify and Implement Personal Actions
You can reduce your carbon footprint by reducing consumption, reusing, recycling, and buying used rather than new...

1. Individual actions:
   a. Prioritize sustainability when making purchasing decisions
   b. Maintain a sustainable standard of living
   c. Reduce food waste
   d. Buy locally-grown foods
   e. Eat a climate-healthy diet and reduce or eliminate beef consumption
   f. Switch to a plant-rich diet
   g. Plant native trees
   h. If you mow a lawn, convert it to a garden. Plant native species that benefit pollinators
   i. Garden – physically reconnect with the earth – the stuff you are made of
      o Grow food
      o Compost
      o Reduce/eliminate use of fertilizers, pesticides, herbicides, etc.

    j.   Buy used rather than new
- Clothes
- Household items
- Cars

    k.   Eliminate single-use plastics
- Use a reusable shopping bag

    l.   Recycle:
- Plastics (1, 2, 5, etc.)
- Glass
- Aluminum
- Metal
- Paper
- Cardboard
- Batteries
- Computers (desktops, tablets, phones, laptops, etc.)

2. Home energy use:

    a.   Use a toaster oven

    b.   Use a microwave to heat water or reheat food instead of a stove or oven

    c.   Replace incandescent light bulbs with energy-efficient LEDs

    d.   Turn off the lights when you leave the room

    e.   Turn off equipment when not in use (computers, TV, lights, etc.)

    f.   Reduce the amount of hot water used
- Wash clothes with cold water
  - Use powdered laundry soap
  - Dry clothes by hanging them instead of using the dryer
  - Use wool dryer balls instead of dryer sheets at a lower dryer temperature
- Take shorter showers

    g.   Only run the dishwasher when full of dishes
- Run the air-dry option

    h. Use a programmable thermostat
       o Adjust for less energy use (cooler in winter, warmer in summer)
       o Reduce heating and cooling when on vacation or at work
    i. Use a smart power strip to eliminate standby energy
    j. Replace outdated appliances with energy-efficient ones:
       o Water heater
       o Heating/Cooling system
       o Refrigerator
       o Oven
       o Washer/Dryer
    k. Insulate doors and windows
    l. Install Solar PV panels
    m. Purchase electricity from a renewable supplier
    n. If you get electricity from solar or wind, convert from gas-powered garden equipment to battery-powered equipment (e.g., mower, weed eater, tiller, etc.)

3. Transportation:
    a. Use an EV, or a plug-in hybrid EV
    b. Ride a bike
    c. Use public transportation
    d. Form carpools
    e. Obey the speed limit
    f. Eliminate unnecessary rapid acceleration
    g. Eliminate unnecessary trips (instead of multiple trips, make one trip to the store)
    h. When possible, walk
    i. Eliminate travel and hold meetings with video teleconferencing software
    j. On long trips, take the train if possible

Carbon Fasting

Have the group members commit to doing a carbon fast and praying for the healing of all creation.

Here are some carbon-fast ideas people have committed to doing:

a. Eliminate unnecessary trips to the store.
    o   Instead of multiple trips, make one trip
    o   Carpool with others
b. Only run the dishwasher when it is full of dishes
c. Reduce the amount of hot water used
    o   Wash clothes in cold water
    o   Take shorter showers
d. Use a toaster oven or microwave instead of an oven or stove
    o   Heat water for tea in the microwave
    o   Reheat food in the microwave
e. Eliminate or greatly reduce the use of the clothes dryer
    o   Air dry clothes
f. Turn off the lights when you leave the room
g. Replace lightbulbs with energy-efficient LEDs
h. Adjust the thermostat for energy efficiency
i. Food Consumption:
    o   Reduce food waste
    o   Buy locally grown food
    o   Reduce beef consumption – eat only locally grown beef
j. Eliminate single-use plastics
k. Bring your reusable shopping bag to the grocery store
l. Rinse, air dry, and reuse Ziplock bags
m. Reduce paper towel usage
n. Recycle:
    o   Paper
    o   Cardboard
    o   General metals
    o   Aluminum cans
    o   Plastics: 1, 2, 5
    o   Batteries
    o   Glass bottles
    o   Computer CPUs

These new activities can move from temporary to normalized new behaviors by putting these actions into regular practice.

Share Personal Experiences and Stories

Invite conversation/study group participants to share their personal stories about ways they have chosen to live more sustainably, take action to help communicate with others about climate change, or work for systemic change in social justice.

For those who have worked to reduce their carbon footprint, share their experiences with others. Learn from each other. Discover what works for you and share your stories.

Since 2018, I have installed solar panels and made several energy efficiency improvements (including replacing the water heater and heat pump). I estimate that my carbon emissions from electricity and heat have been cut from 12.5 to 2.5 metric tons of CO_2 per person annually. I have also reduced my electric bill by \$8,000 and increased my home value by \$15,000. Table 7 summarizes the actions taken that resulted in this CO2 emission reduction.

Table 7: List of Home CO$_2$ Emission Reduction Actions:

Home Electricity and Heating Action Item	CO$_2$ Reduction (Metric tons /year/person)
Install solar PV	6.1
Replace HVAC	1.5
Install programable thermostat	0.6
Replace the water heater	1.0
Launder clothes with cold water, air-dry clothes, and take short showers	0.8
Total	**10.0**

In addition to home energy use, I converted from an internal combustion engine to an all-electric vehicle in 2019. Driving an EV charged by solar PV panels has reduced my carbon emissions from personal transportation from 6.7 to 0.3 metric tons of CO_2 per person annually. In addition to their energy efficiency, EV electric motors and battery metals are recyclable. An electric motor powers EVs and requires much less maintenance than a gasoline-powered car with an internal combustion engine. The net result is tremendous savings on maintenance and gasoline costs. I have saved about $5,000 in fuel and maintenance costs over a five-year period.

In addition, I have written letters and called government officials to advocate for policies that promote a clean energy transition. I have also visited government offices to express my concerns about climate injustice, the impact climate change has on the marginalized, and the need to reduce low-income families' energy costs through energy efficiency programs.

In my journey with Christ, I focus on his purpose for my life, discerned by the gifts he has given and the opportunities he provides to develop and use them. I see the beauty in God's creation. I know my life is connected to the earth and God. When God created the earth, called it good, and placed humans here, God commanded humans to practice responsible dominion and entrusted humans to till and keep the earth. I examine how I harm the earth and how that harms others. I seek to act in ways that promote God's salvation plan - the healing of all creation.

Work for Social and Systemic Change

Gain more information about climate change advocacy groups and take climate action to the systemic level. Become a member of a church climate action community. Learn about public policy advocacy in your local area and join a local citizens' climate lobby group.

1. Networking:
   o Build climate solution networks and become an advocate for climate solutions.
   o Stay updated on current climate policy and politics.
   o Keep updated on the latest climate solutions.
   o Work to start a creation care group in your church.
   o Join an environmental group that does work projects.

2. Public policy:
   o Work through the court system. In a recent court ruling, a Montana judge sided with Montana youth, upholding their claim that a Montana law that failed to consider GHG emissions disobeyed the

state constitution and that the youth have a right to a clean and healthful environment. The court ruled against a Republican climate denier supermajority law benefiting the wealthy fossil fuel industry. Here are some internet accessible public climate policy groups:

- o Climate action network
- o U.S. Climate Action Network
- o Citizens' Climate Lobby group
- o Climate Change Makers
- o Faith Alliance for Climate Action

3. Church resources:

- o *Climate Justice Ministry* is a Mennonite Church USA organization that promotes peace with creation.
- o *The Mennonite Central Committee* works to address climate change, supports a just and sustainable transition away from fossil fuels, advocates for government action, and provides resources to those impacted by climate change. Several resources that advance important topics on climate change, climate action, and climate justice are available on their website.
- o *Anabaptist Climate Collaborative* is an Anabaptist organization that advances thinking and action in faith communities that address climate change.
- o *Creation Justice Ministries* is an ecumenical organization that provides policy action and theological education on the protection, restoration, and just sharing of God's creation.

4. Climate change information:

- o *Our World Data*
- o *The American Association for the Advancement of Science*
- o *The World Resources Institute*
- o *Yale Program on Climate Change Communication* is an organization that helps to advance the science of climate change and provides communication tools on the climate crisis.

5. <u>Other resources</u>:

You may do an internet search for the following resources:

o   Voluntary Gas Tax: to increase the awareness of the amount of gas they use and promote alternative modes of transportation.

o   Difference between landfill and composting: Landfill undergoes anaerobic decomposition, producing CO2+CH4 ~50/50 vs. Composting, which undergoes aerobic decomposition, producing mainly CO_2.

o   The difference between degradable, biodegradable, and compostable.

o   The best laundry detergent powder, liquid, pods, sheets.

o   Prayer Guide: Sam Hamilton-Poore, *Earth Gospel: A Guide to Prayer for God's Creation*, (Nashville, Upper Room Books), 2008.

o   Mennonite Creation Care Network, Assess Your Church.

o   Mennonite Creation Care Network, Every Creature Singing.

o   Lent Carbon Fasting, Stations of the Cross.

Chapter 14. Concluding Thoughts

President Obama said, "We are the first generation to feel the effects of climate change and the last generation to be able to do something about it."[268] Act now to avert the coming climate catastrophe. Make informed decisions. Understand root causes. Do self-examination and spiritual reflection. Join a climate action community. Take corrective action using the best knowledge. Do personal work, get involved in groups, and work with others at the social and systemic level.

If humans continue business as usual, at the current rate of increase in fossil fuel consumption, the Earth will experience a 4.8°C temperature increase by the end of the century compared to pre-industrial times. This temperature rise will dramatically alter the Earth's climate, producing catastrophic human consequences, including famine, forced migration, and conflict.

Climate change is upon us. Human activity has impacted the environment. Actions that produce meaningful change require scientific, economic, political, and moral/justice considerations. Advocates must work to align consumer, business, and government stakeholders to reduce greenhouse gases and achieve the goal of a 1.5 °C global temperature rise. There is still time to avert these catastrophic climate change impacts, but we must act now.

Climate change requires an international, interdisciplinary coalition of academics and public policy experts with professional experience implementing programs that reduce GHG emissions and reverse global warming. We have been doing amazing work and have made meaningful progress toward the clean energy transition. Solar PV, wind, and EVs have made significant advancements on a global scale over the past decade, improving performance, lowering cost, and increasing sustainability. Governments and industries have moved from skepticism to cautious optimism to promoting and supporting the clean energy transition. With all the tremendous global progress, we are still in the beginning stages of this project. We must build upon the current momentum and push for the continued development and implementation of these green technologies. All identified solutions must benefit the environment and society, be economically viable, and provide a world where humans can continue to flourish for future generations.

---

[268] Obama, Barack, Twitter, (September 23, 2014, accessed September 15, 2020.

The Climate is a Common Good

According to Pope Francis, "The climate is a common good, belonging to all, meant for all, [and the responsibility of all] ... We are part of nature, included in it, and thus in constant interaction with it..."[269]

Climate change is a global problem with grave implications. Given the complexity of the ecological crisis and its multiple causes, we need to realize that the solutions will not emerge from just one way of thinking. Humans need to change the way they live; their consumption, use, and disposal behaviors cannot continue as they do today. Humans must reduce their reliance on fossil fuels as an energy source for economic growth and development. The Pope has said that the climate change problem and its impact on life-giving ecosystems have serious social, economic, spiritual, moral, and environmental consequences that will be felt by all peoples of all nations on earth now and for generations to come. God made humans an integral part of the earth's biosphere. Everything in the Earth's biosphere is interconnected and interdependent; God called this good.

The Pope states that we must consider comprehensive solutions that address the human-biosphere interactions, respect human dignity and people's cultures, and incorporate interdisciplinary knowledge and intercultural wisdom. The climate action program must be holistic. We must realize at a deeply personal and social level that we need each other and the life-giving biosphere God created and called good.

Our issues are ecological and social. The solutions we embrace must produce just, sustainable, and meaningful responses to the growing climate and environmental problems.

Current Climate Change Reality

This book has attempted to show that irrational beliefs about climate science, unquestioned assumptions about economic progress, unexamined biases about government actions, and problematic points of view on Biblical creation-care hermeneutics guide harmful environmental behaviors and produce ineffective responses to the climate crisis. Anthropogenic climate change is well documented and adequately demonstrated scientifically. Scientific models predict a catastrophe if humans fail to reduce greenhouse gas emissions. The

---

[269] Pope Francis, *Drawdown: The Most Comprehensive Plan Ever Proposed to Reverse Global Warming*, editor Hawken, Paul, (Penguin: New York, 2017), 190-1.

nature of the disaster that lies ahead has been stated in a 2014 report from the American Association for the Advancement of Science on climate change:

> *"Most climate change projections presume that future changes—greenhouse gas emissions, temperature increases, and effects such as rising sea levels—will happen incrementally. A given amount of emission will lead to a given amount of temperature increase, leading to a given amount of incremental sea level rise. However, the geological record for the climate reflects instances where a relatively small change in one element of climate led to abrupt changes in the system as a whole. In other words, pushing global temperatures past certain thresholds could trigger abrupt, unpredictable, and potentially irreversible changes that have massively disruptive and large-scale impacts. At that point, even if we do not add any additional CO_2 to the atmosphere, potentially unstoppable processes are set in motion. We can think of this as a sudden climate break and steering failure, where the problem and its consequences are no longer something we can control. In climate terms, abrupt change means change occurring over periods as short as decades or even years."*[270]

We all have one thing in common: whether male or female, young or old, liberal or conservative, with great wealth and status, or with few resources and marginalized, we *all* need the earth to survive. We must use and care for the earth's life-giving ecosystems to survive. This is not a zero-sum game; we are *all* in this together. We *all* have a role, a responsibility, and a calling. We must work at the personal, social, and systemic levels to identify and implement corrective actions and make meaningful improvements.

This is a transformational process. It is hard work. It takes holistic dedication that involves using our minds, processing the feelings of our hearts, listening to and reflecting upon the Spirit's wisdom, and making informed decisions that lead to meaningful actions. It requires a learn-as-you-go attitude, a can-do spirit, and an openness to seeing mistakes as learning experiences, not failures. As we do this, we begin to normalize new behaviors that give birth to

---

[270] Malina, Mario, et al., "What We Know: The Reality, Risks, and Response to Climate Change," *The AAAS Climate Science Panel*, American Association for the Advancement of Science, (2014): 6, accessed February 14, 2022.

life-giving relationships, bring results that produce hope, and release inner gratitude for the blessings that come with real life-changing improvements.

We must heal the alienation in our relationships with God, self, others, and the earth. Our concern for the future of humanity and the earth gives energy to the work we have been tasked to do today. By embracing the transformative journey, we discover our gifts, learn how to use them, and uncover the purpose of our lives. By acting in ways that align with our purpose, we make meaning for our lives. We reconcile the alienation and deepen our connection with God, self, others, and the earth. We enter a realm of inner devotion to the purpose we have been called to and outward gratitude for all that God has given us to use and care for. We receive the shalom of understanding that we have been loved so that we can love. Moreover, by knowing the nature of the One who loves and the character of the love, we form a self-giving, other-aware focus, not a self-serving transactional intent; we are made whole when we act responsibly in a loving way.

Part VI

Glossary & Bibliography

Abbreviations

Abbreviation	Explanation
AMOC	Atlantic Meridional Overturning Current
CET	Clean Energy Transition
CFC	Chlorofluorocarbon
CH_4	Methane
CO_2	Carbon dioxide
COP	Conference of the Parties
EIA	Energy Information Administration
EPA	Environmental Protection Agency
EROI	Energy Return on Investment
EV	Electric Vehicle
GHG	Greenhouse Gas
GND	Green New Deal
GW	Gigawatt (billion Watts)
GWh	Gigawatt-hour
HFC	Hydrofluorocarbon
ICE	Internal Combustion Engine
IEA	International Energy Agency
IR	Infrared (heat-producing light)
IRA	Inflation Reduction Act
KW	Kilowatt (thousand Watts)
KWh	Kilowatt-hour
LCOE	Levelized Cost of Electricity
Li-ion	Lithium-ion

MW	Megawatt (million Watts)
MWh	Megawatt-hour
NGCC	Natural Gas Combined Cycle
O_3	Ozone
PHEV	Plug-in Hybrid Electric Vehicle
TW	Terawatt (trillion watts)
TWh	Terawatt-hour

182

Definitions of Common Terms

Term	Meaning
Afforestation	Planting trees where there were no trees before
Alienation	The loss of empathy that leads to separation, estrangement
American Exceptionalism	The belief that the U.S. is exemplary over other nations
Anthropogenic	Human influenced
Atlantic Meridional Overturning Current (AMOC)	Ocean currents circulate water on a long cycle in the Atlantic Ocean, including the Gulf Stream.
Biodegradable	Decomposition of organic matter, such as wood, wool, and cotton, by bacteria into harmless biomass and H_2O plus CO_2 & CH_4 (without oxygen) or CO_2 (with oxygen)
Carbon dioxide (CO_2)	Odorless, colorless, nonflammable gas found in air
Carbon neutrality	This is reached when the amount of CO_2 released into the atmosphere equals the amount removed from the atmosphere by various means.
Chlorofluorocarbon (CFC)	Refrigerant is used in air conditioning and heat pump systems, a potent greenhouse gas.
Clean Energy	Energy that comes from renewable, zero-emission sources without emitting greenhouse gases, such as wind and solar
Clean Energy Transition (CET)	Transition from fossil-fuel-based energy production and consumption, such as coal, oil, and natural gas, to clean energy sources like wind and solar
Climate (also see: Weather)	The totality of weather conditions over decades or more on a global scale
Climate Change (also see: Global warming)	Significant changes in climate patterns, temperatures, and rainfall over long periods
Coastal Virginia Offshore Wind (CVOW)	Project to build large offshore wind turbines to generate 2.6 GW of clean, renewable electrical power for the grid
Compostable	Decomposition of organic matter, such as food scraps, paper, and yard waste, by bacteria in the presence of oxygen and heat into harmless biomass and H_2O plus CO_2
Congress of Parties COP)	United Nations climate change conferences since 1995. COP26's goal is net-zero GHG emissions by 2050.
Ecojustice	Linking environmental concerns with social justice issues
Econationalist	A social movement that uses national interests to define environmental protection policies.
Ecospirituality	An approach to faith that celebrates the human connection to the natural world.

Ecosystems	A collection or community of living organisms that interact with each other and the surrounding environment
Electric Vehicle (EV)	A vehicle that is powered by an electric motor that draws electricity from a rechargeable battery
Energy (Joules)	The capacity to move an object, do work, and produce heat. Energy can exist in the form of sunlight, wind, water movement, electrical, thermal, mechanical, chemical, steam, or nuclear. The unit for energy is Joules.
Energy Return on Investment (EROI)	The ratio of usable energy from an energy source to the amount of energy expended to get that energy source
Environmental Protection Agency (EPA)	Federal agency with the mission to protect human health and the environment
Gigawatt hour (GWh)	One billion Watt-hours
Gigawatt (GW)	One billion Watts
Global Warming (also see: Climate Change)	Rise in average temperatures of oceans, land surface, and atmosphere, caused naturally and by human emissions of greenhouse gases
Greenhouse Gas (GHG)	Gases in the earth's atmosphere, such as CO_2, CH_4, CFCs, and HFCs, that trap heat
Green New Deal (GND)	A plan to transition the U.S. to 100% clean, renewable energy by 2030 using, among other things, a carbon tax
Hydrocarbon	A compound made of hydrogen and carbon, the main component of fossil fuels
Hydrofluorocarbon (HFC)	A refrigerant in air conditioning and heat pump systems is a potent greenhouse gas.
Inflation Reduction Act (IRA)	Federal law that extends clean-energy investment tax credits to provide up to 30% credit for wind, solar, and electric vehicle charging stations
Internal Combustion Engine (ICE)	An engine that does work by internally combusting (burning) a fossil fuel and rubbing the motor
International Energy Agency (IEA)	The agency that works with countries around the world to shape energy policies
Kilowatt-hour KWh	One thousand Watt-hours
Kilowatt KW	One thousand Watts
Levelized Cost of Electricity (LCOE)	The total cost to build and operate a power plant is divided by the amount of electricity generated by the plant over its expected lifetime.
Life-giving	That which sustains, revitalizes, and stimulates growth.
Lithium-ion battery	A rechargeable battery that uses lithium ions (Li+) to generate electrical voltage
Megawatt hour (MWh)	One million Watt-hours
Megawatt (MW)	One million Watts

Metanarrative	An overarching account or interpretation of life's events that assigns root causes and defines a basis for beliefs on how to live
Methane (CH₄)	Odorless, colorless, flammable hydrocarbon gas
Natural Gas Combined Cycle (NGCC)	The heat produced from the combustion of natural gas is used to generate electricity, and heat from the exhaust gas (from the combustion process) is used to make steam, which then generates additional electricity.
Neoplatonic dualism	Religious interpretations of Plato maintain an absolute separation between the material and the spiritual. The material world is transient and trapped in a fallen state, while the spiritual is immortal and perfect.
Net-zero greenhouse gas emissions	Reducing net greenhouse gas emissions in a given year to zero (this includes carbon neutrality)
Ozone (O₃)	Colorless gas with a pungent odor found in the atmosphere, produced naturally or by human activity.
Paris Agreement	A legally binding international agreement on climate change to limit global temperature rise to less than 1.5 °C above pre-industrial levels
Permafrost	A layer of soil and biomass (such as plants and animal remains) that remains permanently frozen is mainly found in polar regions under the Earth's surface, mixed with ice.
Plug-in Hybrid Electric Vehicle (PHEV)	A vehicle that uses a rechargeable battery to power an electric motor and a fossil fuel, such as gasoline, to power an internal combustion engine
Power	The amount of energy produced or consumed per unit of time. In the international system of units, power is joules per second.
PV	Photovoltaic: a material that produces an electric current when exposed to light
Reconciliation	The action of healing relational brokenness, replacing alienation with life-giving relationships that bring about empathy, understanding, intimacy, and vulnerability
Reforestation	Tree planting where the number of trees is decreasing
Renewable Energy	Energy that comes from sources that are naturally replenished, such as biomass, wind, solar, hydroelectric, and geothermal
Solar	Solar energy can be generated by thermal heating using mirrors to concentrate sunlight or by light being converted directly into electricity using solar panels.
Solar PV	Solar PV (Solar-photovoltaic) is solar energy produced when solar panels convert sunlight into electricity.

Self-ascendancy	Personal motivation or action to achieve control or power over one's life and others' lives – seek to become godlike – put human will above God's will
Sustainable Energy	Energy that comes from sources that can be replenished at a rate greater than they are consumed, such as wind, solar, hydroelectric, and geothermal
Terawatt hour (TWh)	One trillion Watt-hours
Terawatt (TW)	One trillion Watts
Watt: a unit of power Kilowatt (kW) Megawatt (MW) Gigawatt (GW) Terawatt (TW)	The amount of energy per second (Joules per second). KW 1,000 watts – one thousand watts MW 1,000,000 watts – one million watts GW 1,000,000,000 watts – one billion watts TW 1,000,000,000,000 watts – one trillion watts
Watt-hour (Wh): unit of energy Kilowatt-hour (KWh) Megawatt-hour (MWh) Gigawatt-hour (GWh) Terawatt-hour (TWh)	Wh the amount of energy (Joules) delivered by producing 1 watt of power (Joules/second) for one hour (3,600 seconds).
Weather (also see: Climate)	Patterns of air circulation, temperature, and precipitation over short periods that are affected by seasonal processes.

Discussion Questions

Chapter 1

Cause and Effect of CO_2 Emissions

Earth is Our Only Home

1. How thick is the Earth's breathable atmosphere?

2. Why is releasing CO2 as an untreated waste stream from burning fossil fuels unsustainable?

How CO_2 Acts as a Greenhouse Gas

3. How does CO_2 act as a greenhouse gas?

4. When did scientists first report the effects of CO_2 on the Earth's atmosphere?

5. What warnings did they give?

The Earth is Rapidly Approaching the Climate Tipping Point

6. What is a climate tipping point?

7. What temperature tipping is the Earth rapidly approaching?

8. How will crossing that tipping point affect the Earth's climate?

Historical CO_2 Levels in the Earth's Atmosphere

9. How much has the average atmospheric CO_2 level increased since 1850?

10. How do we know historical CO_2 levels?

Correlating Atmospheric CO_2 Levels with CO_2 Emissions from Burning Fossil Fuels

11. How many billions of tons of CO_2 are humans putting into the atmosphere yearly from burning fossil fuels?

Impact of CO_2 Levels on Earth's Temperature

12. How much has the global temperature increased since 1850?

13. How do we know historical Earth temperatures?

14. How has this CO_2 increase affected the Earth's temperature?

Impact of Temperature on Earth's Sea Levels

15. What causes the sea level to increase?

16. How much has the sea level risen since 1880?

17. How many people worldwide will be affected by sea-level rise?

18. How much sea level rise is predicted by the end of the century if we do not curb CO_2 emissions?

The Importance of Protecting the Antarctic and Greenland Ice Sheets

19. Where are the primary sources of land-based ice located?

20. How much could the Greenland and Antarctica ice sheets increase the sea level?

21. What cities in the U.S. are most vulnerable to rising sea levels?

Impact of Ice Sheet Melting on the Gulf Stream

22. Why is the Gulf Stream important?

23. How will disrupting the Gulf Stream impact the climate?

There is More than One Type of Greenhouse Gas

24. What different types of greenhouse gases are produced by human activity?

25. What is permafrost? And why is it important?

26. How does CH_4 compare with CO_2 as a greenhouse gas?

Impact of Feedback Cycles on Climate Change

27. What is a climate feedback cycle?

28. How does crossing the climate tipping point affect the pace of climate change?

Chapter 2
Sources and Uses of Fossil Fuels

Which Fossil Fuels Do Humans Use?

1. According to Figure 6, what were the primary sources of CO_2 emissions in 2021? And how much of each fossil fuel type do humans use?

2. How much of each fossil fuel type do humans use?

3. How much CO_2 did human activity release in 2021?

4. What time frame did fossil fuel emissions dramatically increase?

What Do Humans Use Fossil Fuels For?

5. According to Figure 7, what do humans use fossil fuels for?

How Much Total CO_2 Have Humans Produced?

6. According to Figure 8a, what nations have emitted the most CO_2 since 1750?

7. According to Figure 8b, what nations emit the most CO_2 today?

8. According to Figure 8c, what nations have the highest per capita CO_2 emissions?

Fossil Fuel Consumption by Nation

9. According to Table 2, which nations have decreased annual CO2 emissions?

10. According to Figures 9a-9f, which nations have decreased coal consumption?

National Wealth and CO_2 Emissions

11. Is it possible to grow the economy and reduce CO_2 emissions? Explain your answer.

Climate Change Scenarios

12. What is the Paris Agreement's climate goal? What is the COP26 Goal?

13. What is the projected annual CO_2 emission level with no climate policies by the end of the century, and what will the climate impact be?

14. What climate policies can humans take to reduce CO_2 emissions, and what will their impact be?

Chapter 3
Transition to a Clean Energy Economy

Clean Energy Production by Nation
1. Which nation currently generates the most electricity from clean energy sources?

2. What is the fastest-growing clean energy source in the U.S.? Which clean energy source has the greatest electricity-generating capacity?

Power Plant Cost Comparisons (LCOE)
3. What is the Levelized Cost of Electricity (LCOE)?

4. How do the LCOEs of wind and solar compare to other power-generating plants?

Energy Returned on Energy Invested (EROI)
5. What is Energy Returned on Energy Invested (EROI)?

6. How do the EROIs of wind and solar compare to other power-generating plants?

How Do Fossil Fuels, Trees, Solar PVs, Windmills, and Rechargeable Batteries Work?
7. How many pounds of CO_2 are produced by burning one gallon of gasoline?

8. How do trees remove CO_2?

Wind as a Clean-Renewable Energy Source in the U.S.
9. What are the best locations for wind energy in the U.S.?

10. What states have the greatest amount of land-based electricity generation from wind?

Solar PV as a Clean-Renewable Energy Source in the U.S.
11.   What are the best areas for solar PV energy in the U.S.?

12.   Which states currently produce the most electricity from solar PV?

Electric Vehicles for Transportation
13.   How do Electric Vehicles (EVs) benefit the environment versus Internal Combustion Engines (ICEs)?

Conclusion
14.   Can we both reduce CO_2 emissions and grow the economy? (Explain)

15.   Do we have the technology to transition to clean energy in electricity, heat generation, and transportation? (Explain)

Chapter 4
Economics, Populism, and the Cultural Divide

Economics and the Environment

1. What percentage of climate scientists agree that climate change is real and human-caused? What percentage of delegates to COP26 were representatives from the fossil-fuel industry?

2. What benefits does the capitalist economic system provide to society? What are the unwanted consequences? How do "unpriced externalities" affect business plans and the environment?

3. What impact do consumers have on the economy? In what ways may government regulators and free market economists work together to reduce CO_2 emissions?

4. How do international policies affect the spread of clean energy technologies?

5. How does the complexity of climate change make us susceptible to moral corruption?

Populist Opposition

6. What concerns do populists have about climate change? How do they view international cooperation, the role of experts, and the clean energy transition?

7. What actions do populists oppose to address climate change? Which do they support?

8. How do econationalists describe climate change's leading causes and threats?

Climate Change and the Cultural Divide

9. How does one's cultural point of view predict their assessment of climate change?

10. Climate Change Deniers: What arguments are used to refute human-caused climate change?

11. Climate Action Opponents: What reasons are given for not taking climate action?

12. Climate Change and the Religious Divide: What differences in understanding of scripture cause divisions among Christians on Climate Change?

13. Climate Action Advocates: What can advocates do and say to support climate action?

Chapter 5
Climate Action

Eco-overshoot to Sustainability

1. What is Eco-overshoot?

2. How are environmental systems interconnected?

3. Which of the nine planetary boundaries are in eco-overshoot?

4. How can humans improve eco-efficiency?

Is Technology Part of a Just Solution?

5. Can the clean energy transition be accomplished justly and sustainably?

6. Should we live more simply?

7. Is going back to a pre-industrial standard of living feasible?

8. What do you think of managed descent?

9. How does a fossil fuel-based economy's environmental impact compare with a metals-based economy?

Do We Have the Needed Technology, and Can We Get to Net-Zero Emissions in Time?

10. Should we accelerate the clean energy transition?

11. Do we have the technology to meet the COP26 and Paris climate goals?

12. Can we get there on time?

13. What role does political advocacy play?

Chapter 6
A Theology of Creation Care – Section 1

Prophetic Exhortation

1. Do you feel making lifestyle changes to live more sustainably is necessary? (Explain)

2. Does misuse of the Earth's resources create a moral crisis? (Explain)

3. How does Mark 9:43 warn about living unsustainably?

4. What do Genesis 1:28 and 2:15 say about God's command to care for creation?

5. What warning does Revelation 11:18 give to those who destroy the Earth?

The Separation of Heaven from Earth

6. How does the focus on heaven as a place of ultimate personal fulfillment lead to the separation of heaven from earth?

7. What hierarchy does the separation of heaven from earth create?

8. What does the separation hierarchy say about the focus of individual salvation?

9. How does the separation hierarchy affect one's view of creation care?

Distorted Views of Heaven and Earth

10. How may a romanticized view of nature lead to fatalism?

11. How does seeing nature merely as a source of raw materials for human consumption lead to exploitation?

12. How does turning nature into an object of worship lead to idolatry?

13. How does spiritualizing the natural order lead to dualism?

The Seven-fold Barrier

14. How does <u>Greek Philosophy</u> inform the Christian fundamentalist worldview?

15. How does <u>Enlightenment</u> inform the Christian fundamentalist worldview?

16. How does the economic ideology of <u>Capitalism</u> inform the Christian fundamentalist worldview?

17. How does <u>Individualism</u> inform the Christian fundamentalist worldview?

18. How does <u>Nationalism</u> inform the Christian fundamentalist worldview?

19. How does a <u>Dualistic Doctrine of Salvation</u> inform the Christian fundamentalist worldview?

20. How does <u>Premillennial Dispensationalism</u> inform the Christian fundamentalist worldview?

Chapter 7
A Theology of Creation Care – Section 2

Biblical Texts Opposing Dualism

1. What key Biblical texts oppose dualism?

The Four-fold Alienation

2. What is the four-fold alienation?

3. How does the human desire to be like God lead us away from God?

4. How does the separation of heaven from earth distort human understanding of God's salvation plan?

5. What beliefs indicate the separation of heaven from earth?

6. What scripture texts indicate that salvation is the healing of all creation?

The Implication of the Four-fold Alienation

7. What effect does alienation from God have?

8. What effect does self-alienation have?

9. What effect does alienation from others have?

10. What effect does alienation from the Earth have?

Competing Christian Belief Systems

11. How do the dualistic and reconciliation belief systems differ?

Theology Produces Action

12. How do we overcome dualism?

13. What actions do humans need to care for creation meaningfully and effectively?

Chapter 8
Climate Justice

Those Who Contribute the Least Suffer the Most

1. How is climate change a moral and justice issue?

2. Which countries have the highest greenhouse gas emissions, and which countries are most affected by climate change?

Examples of Disproportional Impact on the Marginalized

3. How do emissions from coal-fired power plants impact the most vulnerable?

4. What has been done to reduce harmful emissions from coal-fired power plants?

5. How is the impact of hurricanes a socioeconomic problem?

The Call for a Christian Response

6. Read Isaiah 1:17. How does this text apply to the effects of pollution on the vulnerable ones?

7. Read Luke 1:51-53. How does this text apply to the powerful and the unfortunate ones?

8. Read Matthew 9:12-13. How does this text apply to the comfortable ones?

Chapter 9
Ecospirituality in the Christian Faith Community

The Cain and Abel Dilemma

1. How do self-ascendancy and the zero-sum game affect the Cain and Abel relationship?

2. How do self-ascendancy and the zero-sum game affect the relationship between Western European and North American Indigenous peoples?

Worships and Offerings

3. How do worship and offerings affect our point of view on climate change and the healing of the four-fold alienation?

4. What must we do to move beyond individualism and dualistic salvation to kingdom work in the here and now?

Chapter 10
Looking Inward

Futility and Decay (Romans 8)

1. How do futility and decay impact life on Earth?

2. What causes fatalism, denial, and despair? What are the antidotes?

The Call of Moses (Exodus 3-4)

3. Do you feel uncertain or doubtful that you can make a difference?

4. What excuses did Moses give to God? How did God respond to Moses' doubts?

5. What is God calling you to do? How do you respond to God's call?

Chapter 11
Seeing Self and Others Clearly

Transitioning from Divisive Debates to Constructive Conversations

1. What role does the facilitator have in establishing constructive conversations?

2. What can the facilitator do to promote healthy conversations that produce critical reflections and informed action?

How to have Constructive Conversations about Climate

3. How does becoming fixated on proving a point hinder constructive conversation?

4. Why is it necessary to process feelings to make connections?

Bring People Together and Set Some Goals

5. How does setting conversation goals help keep people from being defensive or dismissive?

Understand Why Climate Science is Rejected

6. Why do some people reject climate science?

Connect Who You Are to Why You Care

7. What do we have in common, and what shared values do we have that reveal shared concerns about climate change?

Chapter 12
Use Your Heart, Mind, Body, and Spirit

Welcome Divine Grace
1. What have you learned about climate change?

2. How are you feeling about climate change?

3. What self-examination and critical reflection have brought new awareness that guides you to informed action?

Prayer and Worship
4. How have your prayers and worship helped you to remain rooted in hope and action?

Personal Action and Systemic Change
5. How are you willing to change your consumer choices?

6. What personal actions can you take?

7. What systemic changes do you feel called to work on?

Chapter 13
Use Best Knowledge and Take Informed Action

Practice Strategic Thinking

1. What can you start doing to reduce your carbon footprint now?

2. How can you reduce home energy use?

3. How can you reduce your transportation carbon emissions?

4. What changes can you make to live more sustainably?

5. What stories can you share about your personal experiences of living more sustainably?

Bibliography and References

A.G.C. Graham, et al., "Rapid retreat of Thwaites Glacier in the pre-satellite era," *Nature Geoscience* 15, (September 5, 2022): 706–713. accessed September 15, 2022.

Adapted from Francis Weller, *The Wild Edge of Sorrow,* (Berkeley: North Atlantic Books, 2015).

Alder, Ben, "'Doomsday glacier' is melting faster than thought, study finds," *Yahoo!News*, (September 7, 2022). accessed September 15, 2022.

Alexander, Tammy, Editor, "U.S. Policy Towards DR Congo," *Peace and Justice Journal*, (Mennonite Central Committee U.S. Vol LVI, No. 3, Fall/Winter 2024), accessed October 6, 2024.

Alternative Fuels Data Center, "Electric Vehicle Charging Stations," U.S. Department of Energy, accessed September 17, 2024.

Amadeo, Kimberly, "Components of GDP Explained: Four Critical Drivers of America's Economy," *The Balance*, (updated January 18, 2022) accessed September 15, 2022.

Amnesty International, "Powering Change or Business as Usual," Amnesty International, (Updated September 12, 2023), accessed October 17, 2023.

Andre, Peter, et al., "Globally Representative Evidence on the Actual and Perceived Support for Climate Action," *Nature Climate Change*, 14, (March 2024): 253-259.

Argonne National Laboratory, "Light Duty Electric Drive Vehicles Monthly Sales Updates," *U.S. Department of Energy*, (updated September 2023), accessed October 17, 2023.

Armstrong McKay, David I., et al., "Exceeding 1.5 C Global Warming Could Trigger Multiple Climate Tipping Points," *Science*, 377 Issue 6611, (September 9, 2022).

Arnold, Bill T., Genesis, (New York: Cambridge University Press, 2009).

Arocho-Esteves, Juno, "Failure to protect creation will mean facing 'God's judgment,' Pope Francis says," *EarthBeat*, (Nov. 11, 2021), accessed Dec. 9, 2021.

Balmer, Randall, *The Making of Evangelism: From Revelation to Politics and Beyond*, (Waco, Baylor University Press, 2010).

Bartels, Meghan, "Humans Have Crossed 6 of 9 'Planetary Boundaries," *Scientific American,* (posted September 13, 2023), accessed October 17, 2023.

Bauckham, Richard, *Bible and Mission: Christian Witness in a Postmodern World* (UK: Baker, 2005).

Bauer, Bruce, "How Tree Rings Tell Time and Climate History," *National Ocean and Atmospheric Administration*, (11/29/2018), accessed April 6, 2023).

Bauer, Michael, "The State of Global Air/2018: A Special Report on Global Exposure to Air Pollution and its Disease Burden," *HEI and IHME* (2018): 1, (accessed May 25, 2023.

Block, Samuel, "Mining Energy-Transition Metals: National Aims, Local Conflicts," *ESG Research*, (June 3, 2012), accessed October 18, 2022.

Box., J. E., Hubbard, A., Bahr, D.B., et al. "Greenland ice sheet climate disequilibrium and committed sea-level rise," Nature Climate Change 12 (August 29, 2022), 808, 808-813, accessed September .29, 2022

Bush, Evan, "In Search of 10 billion Missing Snow Crabs, Scientists Eye Marine Heat Waves," *NBC News*, (published October 20, 2023) accessed October 21, 2023.

Carbajales-Dale, Michael, "Energy Return on Investment," *Science Direct*, accessed August 8, 2023.

Catholic Climate Covenant, "Pope Francis: A Prayer for Our Earth," Franciscan Monastery, (Jul. 30, 2024), accessed Aug. 8, 2024.

Chapman, Becky, "How Does a Lithium-Ion Battery Work?" *Let's Talk Science*, (September 23, 2019), accessed 11/08/2023.

Claiborne, Shane, et al., *Common Prayer: A Liturgy for Ordinary Radicals*, (Grand Rapids: Zondervan, 2010).

Climate Action Tracker (2022), The CAT Thermometer, November 2022. Copyright © 2022 by Climate Analytics and New Climate Institute. All rights reserved.

Clymer, Donald and Clymer-Landis, Sharon, *The Spacious Heart: Room for Spiritual Awakening*, (Harrisonburg, Herald Press, 2014).

Comay, Laura B., and Clark, Corrie E., "Offshore Wind Energy: Federal Leasing, Permitting, Deployment, and Revenues," Congressional Research Service, (updated December 7, 2021), accessed September 15, 2022.

Cozzi, Laura, "Sustainable Recovery," *International Energy Agency*," (revised July 2020), accessed October 28, 2023.

De Menocal, Peter, "A Climate Science Refresher," *Columbia Climate School*, (March 1, 2017), accessed December 4, 2021.

Diamond, Howard, "Science and Information for a Climate Smart Nation," *NOAA*, (July 8, 2021), accessed December 4, 2021.

Dominion Energy, CVOW Project, (copyright 2023), accessed October 28, 2023.

Dorman, John L., "No Lectures on How to 'Destroy' Economy Over Climate Change," *Business Insider*, Jul 23, 2022, accessed August 1, 2022.

Dorn, Sara, "DeSantis Under Fire for Rejecting Millions in Home Energy Funding from Biden's Inflation Reduction Act," *Forbes*, (updated August 30, 2023) accessed August 31, 2023.

Dula, Peter, "Anabaptist Environmental Ethics: A Review Essay," *Mennonite Quarterly Review* 94, no. 1 (January 2020).

Dula, Peter, Eastern Mennonite Seminary Chapel presentation, *Eco Theology*, April 19, 2022.

Eberhart, Timothy R., "New Birth toward a New Earth Regenerative Mission for Planetary Renewal," Missio Dei and the United States: toward a Faithful United Methodist Witness, Kathryn M. Armistead, editor, (USA: The United Methodist Board of Higher Education and Ministry, 2018).

Educational Broadcasting Corporation, "Slavery and the Making of America," *Public Broadcasting Service*, (published 2004) accessed November 4, 2023.

EIA, "Solar and Wind to Lead Growth of U.S. Power Generation for the Next Two Years," *U.S. Energy Information Administration*, (January 16, 2024), accessed June 27, 2024.

EIA, "Table 1.2: Summary Statistics for the U.S. 2012-2022," *U.S. Energy Information Administration*, (October 19, 2023), accessed June 27., 2024

Ember's Yearly Electricity Data, "Electricity Generation from Low-Carbon Sources," *Our World Data*, (April 2022), accessed September 15, 2022.

Encyclopedia Virginia, "Laws of Virginia," *Library of Virginia*, (published 2020), accessed November 4, 2023.

EPA, "Electric and Plug-in Hybrid Electric Vehicles," *United States Environmental Protection Agency*, (updated August 28, 2023), accessed August 29, 2023.

EPA, "Electric Vehicle Myths," *United States Environmental Protection Agency*, (updated August 28, 2023), accessed August 29, 2023.

EPA, "Greenhouse Gas Emissions," *United States Environmental Protection Agency*, (updated April 2023), accessed August 8, 2023.

EPA, "Improving Recycling and Management of Renewable Energy Wastes: Universal Waste Regulations for Solar Panels and Lithium Batteries," *United States Environmental Protection Agency*, (updated December 14, 2023), accessed June 27, 2024.

EPA, "Renewable Energy Fact Sheet: Wind Turbines," *United States Environmental Protection Agency* (updated August 2013), accessed August 10, 2023.

Fernandez, Lucia, "Leading States in Solar Photovoltaic Capacity in 2022," *Statista: Environment and Energy*, (October 20, 2023) accessed June 27, 2024.

Ferreira, Fernanda, and Odell, Scott, "How Does the Environmental Impact of Mining for Clean Energy Metals Compare to Mining for Coal, Oil, and Gas?" *MIT Climate Portal*, (published May 8, 2023) accessed October 28, 2023.

Foley, Jonathan, "Climate Solutions 101: Putting it All Together," *Project Drawdown*, (published), accessed October 20, 2023.

Foote, Eunice, "Circumstances Affecting the Heat of the Sun's Rays," *The American Journal of Science and Arts*, vol. XXII, (November, 1856): 383. It may be found on Google Books.

Freeman, Jody, Guzman, Andrew, "Climate Change and U.S. Interests," *Columbia Law Review*, 109, no. 6 (Oct. 2009).

Gardiner, Stephen M., "A Perfect Moral Storm: Climate Change, International Ethics and the Problem of Moral Corruption," *Environmental Values*, 15, no. 3 (August 2006).

Garen, David C., "Mennonite Values in a Warming World," *Anabaptist World*, (May 20, 2019), accessed) March 4, 2024.

Global Energy Monitor, "Coal Plants Near Residential Areas," *GEM wiki* (April 30, 2021) accessed May 25, 2023.

Groome, Thomas H., *Christian Religious Education*. (San Francisco: Jossey-Bass Inc., 1999).

Guterres, Antonio, "UN Chief calls for action to put out '5-alarm global fire'", *UN Climate Action* (January 2022), accessed February 10, 2022).

Hall, Charles A., et al., "EROI for Different Fuels and the implications for society," *Energy Policy*, 64 (2014) 141-152, accessed August 8, 2023.

Hamilton-Poore, Sam, *Earth Gospel: A Guide to Prayer for God's Creation*, (Nashville: Upper Room Books, 2008).

Harvey, Chelsea, "Oceans Are Warming Faster Than Predicted," Scientific American, (January 11, 2019), accessed September 15, 2022.

Hawken, Paul, *Drawdown: The Most Comprehensive Plan Ever Proposed to Reverse Global Warming*, editor Hawken, Paul, (Penguin: New York, 2017).

Hayhoe, Katharine, *Saving Us: A Climate Scientist's Case for Hope and Healing in a Divided World,* (New York: Simon & Schuster, 2021).

Hayhoe, Katherine, "The Evangelical Christians Taking on Climate Change," *How to Save a Planet*, Podcast Hosted by: Alex Blumberg (Dec. 9, 2021), accessed 1/15/2022.

Heltzel, Paul, "Global Warming Right Before Your Eyes," *Seeker*, (Published 5/14/2014), accessed October 14, 2023

Hoffs, Charlie, "Challenges and Opportunities in Mining Materials for Energy Storage Lithium-ion Batteries," Union of Concerned Scientists, (updated December 22, 2022), accessed October 17, 2023.

Homer-Dixon, Thomas F., "On the Threshold: Environmental Changes as Causes of Acute Conflict," *International Security*, 16, no. 2 (Fall 1991).

"Horizontal-Axis Wind Turbine (HAWT) Working Principle," *Electrical Academia*, accessed August 10, 2023.

Houreld, Katharine, "Clean Cars, Hidden Toll," *The Washington Post* (Updated August 4, 2023) accessed October 17, 2023.

Inman, Mason, "How to Measure the true Cost of Fossil Fuels," *Scientific American: Environment*, (updated April 1, 2013), accessed August 8, 2023.

Jackson, Margaret, and Strauss, Zachary, "Raising US Climate Ambition in Advance of COP26: An Economic and National Security Imperative," Atlantic Council (2021).

Jansen, Wendy, "About Us," Burning Bush Forest Church, (Updated December 2023), accessed December 12, 2023.

Jayawardhan, Shweta, "Vulnerability and Climate Change Induces Human Displacement," *Consilience: The Journal of Sustainable Development*, 17, no. 1 (2017).

Kauffman, Bradley, General Editor, *Voices Together*, (Harrisonburg, MennoMedia: 2020).

Kottasova, Ivana, and Picheta, Rob, "Greta Thunberg slams COP26 as a 'failure' at youth protest in Glasgow," *CNN* (November 5, 2021), accessed December 6, 2021.

Krause, Eleanor, and Reeves, Richard V., "Hurricanes Hit the Poor Hardest," *Brookings Institution* (September 18, 2017), accessed May 25, 2023.

Kauffman, Bradley, General Editor, *Voices Together*, (Harrisonburg, MennoMedia: 2020).

Larson, Gary, "Animal Waste Management," *Far Side*, Pinterest, accessed October 2023.

Larter, Robert, "Doomsday Glacier 'Holding on by Its Fingernails' – Spine-Chilling Retreat Could Raise Sea Levels by 10 Feet," *SciTechDaily*, University of South Florida, (September 7, 2022), accessed September 15, 2022.

Leiserowitz, Anthony, et al., "Global Warming's Six Americas," Yale Program on Climate Communication, (updated 2023), accessed November 12, 2023.

Lelieveld, J., et al., "Effects of fossil fuel and total anthropogenic emission removal on public health and climate," *Proceedings in the National Academy of Sciences*, April 9, 2019, 116 (15).

Long, Heather, "Where Harvey is hitting the hardest, 80% lack flood insurance," *Washington Post,* (August 29, 2017), accessed May 25, 2023.

Maddox, Randy, "John Wesley's Precedent for Theological Engagement with the Natural Sciences," *Wesleyan Theological Journal* 44, no. 1 (Spring 2009).

Malina, Mario, et al., "What We Know: The Reality, Risks, and Response to Climate Change," *The AAAS Climate Science Panel*, American Association for the Advancement of Science, (2014): 6, accessed February 14, 2022.

Marcacci, Silvio, "Renewable Energy Prices Hit Record Lows: How Can Utilities Benefit from Unstoppable Solar and Wind?" *Forbes: Business-Energy*, (April 14, 2022), accessed June 27, 2024.

Mark, Joshua J., "Virginian Slave Laws and Colonial Development of Colonial American Slavery," *World History Encyclopedia*, (published April 27, 2021), accessed November 4, 2023.

McDowell, Tim, "Ron DeSantis's Climate Contradictions," *SEMAFOR*, (updated April 7, 2023), accessed August 31, 2023.

Martinez, Chris, et al., "These Fossil Fuel Industry Tactics Are Fueling Democratic Backsliding," *Center for American Progress (CAP)*, (Dec. 5, 2023), accessed June 17, 2024.

Meadows, Donella, et al., *Limits to Growth: The 30-Year Update*, (White River Junction: Chelsea Green Publishing, 2004).

Milman, Oliver, "Climate denial is waning on the right. What's replacing it might be just as scary," *The Guardian*, (Nov. 21, 2021), accessed 12/6/2021.

Milman, Oliver, "There's lithium in them thar hills – but fears grow over US 'white gold' boom," *The Guardian*, (October 18, 2022), accessed October 18, 2022.

Mooney, Chris, "Greenland ice sheet set to raise sea levels by nearly a foot, study finds," *The Washington Post: Climate and Environment*, (August 29, 2022), accessed September 29, 2022.

Naomi Klein, *This Changes Everything*, (New York: Simon & Schuster, 2014).

NASA, "The sun shines above the earth's horizon as the International Space Station orbited 264 miles above the Canadian Province of Quebec," *Phys.org* (December 21, 2021), accessed April 21, 2022.

National Ocean Service, "What is the Atlantic Meridional Overturning Current?" *National Oceanic and Atmospheric Administration*, (updated 1/20/2023), accessed October 14, 2023.

NOAA, "Sea Level Rise Viewer," *National Oceanic and Atmospheric Administration*, (updated June 2023), accessed October 14, 2023.

Nordhaus, William D., and Yang, Zili, "Regional Dynamic General-Equilibrium Model of Alternative Climate-Change Strategies, *The American Economic Review*, 86, no. 4 (Sep. 1996).

Obama, Barack, Twitter, (September 23, 2014, accessed September 15, 2020.

Osmanbasic, Edis, "Solid-State EV Batteries are closer than you think," engineering.com, (February 22, 2024) accessed 8/2/2024.

Park, Jonathan T., "Climate Change and Capitalism," *Consilience: The Journal of Sustainable Development*, 14, no. 2 (2015).

Pope Francis, *Drawdown: The most comprehensive plan ever proposed to reverse global warming*, editor Hawken, Paul, (Penguin: New York, 2017).

Pratt, Fran, "A Collection for Litanies from Fran Pratt," Litany for Worship (August 10, 2016) accessed November 20, 2024.

Pries, Betty, *The Space Between Us: Conversations about Transforming Conflict*, (Harrisonburg, Herald Press, 2021).

Ramanujan, Krishna, "More than 99.9% of studies agree: Human-caused climate change," *Cornell Chronicle*, (updated October 19, 2021) accessed August 12, 2023.

Real Gross Domestic Product Per Capita, *FRED Economic Data*, (updated Jan. 25, 2024), accessed 2/10/2024.

Reidmiller, David, et al., *The Climate Report: The National Climate Assessment – Impacts, Risks, and Adaptation in the United States*, (Brooklyn: Melville House, 2018).

Richardson, Katherine, et al., "Earth Beyond Six of Nine Planetary Boundaries," *Science Advances*, (published September 13, 2023), accessed October 18, 2023.

Ritchie, Hannah, and Roser, Max, "CO2 Emissions," *Our World Data*, (2022) accessed September 15, 2022.

Ritchie, Hannah, and Roser, Max, "Future Greenhouse Gas Emissions Scenarios," *Our World Data*, (April 2022), accessed September 15, 2022.

Ritchie, Hannah, Roser, Max, and Rosado, Pablo, "CO_2 Concentrations in the Atmosphere," *Our World Data*, (August 2022), accessed September 15, 2022.

Ritchie, Hannah, Roser, Max, and Rosado, Pablo, "CO_2 Emissions," *Our World Data*, (April 2022), accessed September 15, 2022.

Ritchie, Hannah, Roser, Max, and Rosado, Pablo.

Rockstrom, Johan, et al., "Planetary Boundaries: Exploring the Safe Operating Space for Humanity," Ecology & Society, vol 14(2), 2009.

Rydelnik, Michael, and Vanlaningham, Michael, *The Moody Bible Commentary: A One-Volume Commentary on the Whole Bible by the Faculty of Moody Bible Institute*, (Chicago, Moody Publishers, 2014).

Ryrie, Charles Caldwell, *The Ryrie Study Bible: Expanded Edition, New International Version*, (Chicago: Moody Press, 1994).

"Solar Inverters," *Energy Saving Trust*, (updated 2024), accessed 2/10/2024.

Samson, J., Berteaux, D., and McGill, B.J., Humphries, M.M., "Geographic disparities and moral hazards in the predicted impacts of climate change on human populations," *Global Ecology and Biogeography: A Journal of Microbiology*, (Feb. 17, 2011), accessed 1/29/2024.

Schaeffer, Francis, A., *Pollution and Death of Man: The Christian View of Ecology*, (Wheaton, Tyndale House, 1970).

Schneider, Conrad, and Banks, Jonathan, "The Toll from Coal: An Updated Assessment of Death and Disease from America's Dirtiest Energy Source," *Clean Air Task Force*, (September 2010), accessed May 25, 2023.

Schutte, Daniel, "Here I am Lord," *Voices Together*, Kauffman, Bradley, General Editor, (Harrisonburg, MennoMedia: 2020), 545.

Seibert, Megan K., Rees, William E., "Through the Eye of a Needle: An Eco-Heterodox Perspective on the Renewable Energy Transition," *Energies* 14, no. 15: 4508, p. 2.

Sengupta, M., Y. Xie, A. Lopez, A. Habte, G. Maclaurin, and J. Shelby. "The National Solar Radiation Data Base (NSRDB)." Renewable and Sustainable Energy Reviews 89 (June 2018): 51-60.

Shepardson, David, "U.S. Senator to hold EV battery hearing if GOP takes control," *Reuters*, (October 19, 2022), accessed October 20, 2022.

Snyder, Howard A., *Salvation Means Creation Healed: The Ecology of Sin and Grace* (Eugene: Cascade Books, 2011).

"Solar Cell Construction and Working Principle," *Electrical Engineering 123*, Accessed October 20, 2023.

Stearns, Richard, *The Hole in Our Gospel* (Nashville: Thomas Nelson, 2009).

Stockholm Resilience Centre, "Planetary Boundaries," Stockholm University, (published September 2023), accessed October 17, 2023.

Strauss, Benjamin, "What Does the Earth Look Like with 10 Feet of Sea Level Rise?" *Climate Central*, (May 13, 2014), accessed October 14, 2023.

Sweet, W. V., et al., "Climate Change Indicators: Sea Level," *United States Environmental Protection Agency*, (updated July 2022), accessed September 15, 2022.

Sweet, W.V., et al., "Global and Regional Sea Level Rise Scenarios for the United States," *National Oceanic and Atmospheric Administration*, (February 2022), accessed October 20, 2022.

Sweet, W.V., et al., "Global and Regional Sea Level Rise Scenarios for the United States," *National Oceanic and Atmospheric Administration*, (February 2022), accessed October 20, 2022.

Szuwalski, Cody, S., et al., "The Collapse of Eastern Bering Sea Snow Crab," Science, vol 382 (6668), (published October 19, 2023), accessed October 20, 2023.

The Associated Press, "At COP26, nations strike a climate deal with coal compromise," *National Public Radio*, (November 13, 2021), accessed December 6, 2021.

The National Resource Defense Council, "Cost of Building Power Plants in Your State," accessed September 15, 2022.

The Office of Energy Efficiency & Renewable Energy, "Wind Energy Maps and Data," *Wind Energy Technologies Office – Wind Exchange*, accessed September 15, 2022.

The U.S. Energy Information Administration, "Wind explained: Where wind power is harnessed," *Environmental Impact Assessment (EIA)*, (updated April 20, 2023), accessed August 15, 2023..

The U.S. Energy Information Administration, "Wind explained: Where wind power is harnessed," *Environmental Impact Assessment (EIA)*, (updated March 30, 2022), accessed September 15, 2022..

The U.S. Energy Information Administration.

Thomas, Steve, "My Prayer: God Save the Earth," *Anabaptist World*, (August 26, 2022), accessed September 2022.

UN Environment Programme, "Protecting Whales to Protect the Planet," *Ocean and Coasts*, (14 Oct. 2019), accessed November 19, 2023..

Vikram Solar, "EROI of Solar Energy," (Updated April 21, 2016), accessed August 8, 2023.

Wiebe, Joseph R., "Reassessing Mennonite Environmentalism through Settler-Colonialism: Political Deficiencies, Historical Omissions, and Indigenous responses," The Mennonite Quarterly Review, 96, no. 3, (July 2022): 358.

Wiki, "Greenhouse Gas," Wikipedia, (updated August 2023), accessed August 17, 2023.

White, Jonathan, "What Makes Climate Change a Populist Issue," *Grantham Research Institute on Climate Change and the Environment*, (September 14, 2023), accessed June 20, 2024.

Wright, N.T., *Surprised by Hope: Rethinking Heaven, the Resurrection, and the Mission of the Church*, (New York, HarperOne, 2008).

Wulf, Andrea, "Alexander von Humbolt," Drawdown: The most comprehensive plan ever proposed to reverse global warming," editor Hawken, Paul, (Penguin: New York, 2017).

Yu, Mingliang, et al., "Unlocking Iron Metal as a Cathode for Sustainable Li-ion Batteries by an Anion Solid Solution," *Science Advances*, Vol. 10, Issue 21, (May 23, 2024), accessed 7/20/2024.

www.ingramcontent.com/pod-product-compliance
Lightning Source LLC
Chambersburg PA
CBHW060137130626
46556CB00006B/2388